D0423239

Technology and Planned Organizational Change

James C. Taylor *CHAPMAN* *1937-*

Center for Research on Utilization of Scientific Knowledge

Institute for Social Research
The University of Michigan
Ann Arbor, Michigan

ISR Code Number: 2978-71

Library of Congress Catalog Card Number: 78-161549

ISBN 87944-002-3 (Paperbound)
ISBN 87944-003-1 (Clothbound)

Printed by Braun-Brumfield, Inc.
Ann Arbor, Michigan

Manufactured in the United States of America

PREFACE

The data for this monograph were taken from longitudinal, comparative surveys of industrial organizations, conducted by the Center for Research on Utilization of Scientific Knowledge (CRUSK) between 1967 and 1969. The purpose of the present study was to test several specific notions advanced by socio-technical theorists regarding the ability of technical systems to facilitate planned social change efforts directed toward greater worker participation. The socio-technical approach to industrial organizations provides an interesting and plausible theory which, in conceptual strength, is much beyond the main body of literature linking organizational behavior and technology. Although socio-technical theory is based on empirical study, the data are relatively limited in terms of method used and design applied in their collection. Therefore, an independent test of some predictions from socio-technical theory using longitudinal and quantitative survey methodology was felt to be useful.

Aspects of particular interest in the study were the specific predictions within the general hypothesis. They fell into three categories: influence of supervisory leadership *vis-a-vis* technology in initiating planned change; self-maintenance of peer leadership in planned change; and the precursors to and reinforcing effects of changed attitudes on subsequent behaviors.

One of the main concerns in this study was the assessment of causal priorities among the variables examined. The hypotheses and specific predictions were originally couched in terms of a direct comparison of the effects of technology with the other independent variables. Following these hypotheses, the original data analysis plan involved the use of multiple regression techniques (Path Analysis) in assessing causality. Once collected, however, the data revealed clear deviation from the conditions of linearity and additivity, which are inviolate when using Path Analysis. Tests of the data, therefore, took the more simple form of comparisons of mean scores and correlations, controlling for high and low technological sophistication.

The monograph itself is in two parts: the first part, the original study, was completed in 1969; the second part, the replication, utilized data collected and analysed a year later. It is encouraging to note the degree to which the original and replication data agree, given the differences in sample and data collection design. The replication not only lends credence to the original results, but it provides the opportunity to further examine the role of first-line supervision in planned change.

Although the study began by using a model of technological constraints operating directly on individual behavior as the sole-cause-thesis of technological effect, subsequent results revealed that technological effects on organizational behavior seem to be indirect as well—operating through an interdependence of system subunits.

The financial support for this study came from three sources within the Institute for Social Research. Part of the support came from monies obtained through an Office of Naval Research Grant (Contract NOOO 14-67-A-0181-0013). Part came from CRUSK research overhead funds, and part came from a basic research fund in the Business and Industry Group in CRUSK. I record my gratitude to the persons responsible for the sponsorship: Rensis Likert, Director, Institute for Social Research; Floyd C. Mann, Director, Center for Research on Utilization of Scientific Knowledge; and David G. Bowers, Program Director, Business and Industry Group.

I gratefully acknowledge the substantive assistance of my doctoral committee—Frank M. Andrews, David G. Bowers, Edward O. Laumann, Donald N. Michael, and its chairman, Stanley E. Seashore. Each member provided uniquely appropriate assistance, counsel, and support when most needed.

Thanks are also due to many others who in some way helped to make the resultant study a reality. My colleagues in the Business and Industry Group were of indispensible assistance in too many ways to enumerate. These people include Jon Barrett, Steven Iman, Mark Frohman, Robert Norman, and John Thomas. The persons in the organizations studied who participated in the evaluative tasks were most cooperative and helpful. Although remaining nameless, their efforts are nonetheless greatly appreciated. Able assistance in data analysis and processing came from Nancy Batzer, Mauricio Font, George Mylonas, Carol Shirley, and Edith Wessner. Peggyann Chevalier and Elaine Watson were responsible for the many typing tasks in earlier drafts and manuscripts. CRUSK Editor Joyce L. Kornbluh prepared the monograph for publication.

Indirect, but no less important contributions were provided by L. Richard Hoffman, presently of The University of Chicago School of Business, and Daniel Katz of The University of Michigan Department of Psychology, who as teachers provided molecular as well as molar inputs to the substance and spirit of this monograph.

Finally, to my wife Linda, I owe more than I can say. She provided the support, patience, love (not to mention help in draft editing, and typing), which made it possible.

Such merits as this study may have are due to the encouragement and criticism of my friends and colleagues. For such errors and obscurities as may remain, I bear, of course, the sole responsibility.

James C. Taylor

TABLE OF CONTENTS

Page

TABLE OF CONTENTS

LIST OF TABLES

LIST OF FIGURES

LIST OF APPENDICES

Page

1

THE PROBLEM: TECHNOLOGY AND SOCIAL CHANGE

The area of industrial technology innovation, and its relationship with social system change, is confused and unsystematic. Little arm-chair speculation and theory building has been undertaken, little more research has been done, and of that, most studies are unrelated to any programmatic effort, are purely empirical, and are freighted with methodological problems. The body of accumulated knowledge, however, is large enough so that certain apparent commonalities with regard to the interaction between technical and social systems can be tested. The present research proposes to test the notion, originally advanced by people at England's Tavistock Institute, that work group resistance to planned social system change will be lower, and acceptance higher, in subunits with initially high technology. It is postulated that these facilitating effects come about as a function of the amount of situational constraint applied by the technology, and the degree to which it can be considered an "unfreezing" event in the lifespace of the work group.

Relationships between work group technology and subsequent member perceptions of supervisory and peer group leadership, work group activities, and satisfaction with the work group will be examined. Data used are from a longitudinal study of two organizations, white-collar as well as blue-collar, with a combined total of nearly 300 work groups. Questionnaire mean scores by work group are the primary type of data, and correlations are the primary statistic for measuring relationships.

Literature Review

Technology, or the techniques involved and the principles invoked in bringing about change in materials toward desired ends or products, has been regarded as an important determinant of organizational functioning. Although many organizational theorists prefer to ignore technology, there is a loose grouping of social scientists and philosophers which emphasizes the causal

1

importance of this variable. Apparent pluralistic ignorance exists among these social scientists, however, for despite their affinity for similar predictions, they come from various disciplines and seem not to be aware of one another's work (Pugh, 1966). This is not to say that they speak of technological determination as the one best way to organize work, but that they emphasize the importance of technology as one determinant among many which is frequently overlooked by others in the field of organizational theory.

During the past 15 years, a body of literature has accumulated around this topic of individual and organizational response to increased automation and other forms of technological change. However, little if any consistent scheme of classification has been applied enabling the conclusions of one writer to support or refute the conclusions of another. Much of the extant literature is conjecture and speculation—little evidence is available for most of it. A recent review of the empirical literature (Taylor, 1968) reveals that there are certain consistencies in many of the findings included in this collection of studies, and that these consistencies could be best assessed by separating the studies into those undertaken in blue-collar industrial firms from those done in white-collar organizations, or white-collar subsystems in industrial firms. The present condensed review, presented below, will rely on this separation, by type of work undertaken by the industries studied, and by schemes used to categorize technology which have been developed by the researchers undertaking the studies.

Classification of Technology

A review of the schemes cataloging technology reveals that there seems to be little agreement on what basis to start. Basic thrust of the various systems ranges from Woodward's form of production (1965) through the worker, or role-oriented systems of Blauner (1964) and Touraine (1962); the systems based on standardization of techniques and/or materials (Thompson, 1967; Perrow, 1967) and on the standardization of product (Harvey, 1968); to the very technical or hardware-oriented approaches of Bright (1958) and Faunce (1968). Each of these theorists provides limited, usually blue-collar industry-based evidence that cataloging technology their way allows differentiation within the organizational structure, nature of jobs and work, and employee attitudes. It is not only automated factory equipment, however, which accounts for the rise in modern technology, but computers and office automation as well. An aspect of generality required but rarely found in a system of classification of technology would be the degree to which the scheme includes advances in computer technology or technology in white-collar organizations as well as technology in manufacturing and industrial operations.

Of the classification schemes mentioned above, few of them have direct applicability for white-collar automation. The systems proposed by Blauner, Bright, Faunce, Harvey, Touraine, and Woodward are based either on the technical hardware and its materials handling, or on the dimensionality of the product (e.g., the products of continuous flow chemical plants versus those of durable goods fabrication). The schemes presented by Perrow and Thompson, on the other hand, tend to place more emphasis on techniques based on the complexity of the raw material (as perceived by the influence source or operator). These latter authors provide more general schemes which would include classification of white-collar organizations with greater ease. This is not to say that the more industrially-oriented theories could not be extended for white-collar purposes, but the importance of the fine distinctions, for example, between mass production and continuous process in white-collar work seem less useful than in blue-collar work. The machine-related human behaviors specified in the manufacturing-industrial theories have limited application in other organizations. Computer operators, as the white-collar analog to the automatic machine operators in industry, presently constitute such a small proportion of white-collar labor as to make the extension almost meaningless.

A useful and generally applicable system of classifying technology might involve not only an assessment of the sophistication of machines (e.g., Bright) and an assessment of the sophistication of the raw materials (e.g., Perrow), but the assessment of the complexity of task attributes including prescribed and discretionary activities as well (Turner and Lawrence, 1965). *The best definition of technology under such a scheme would be: a set of principles and techniques useful to bring about change toward desired ends.* Although very broad, this definition seems at once to satisfy the general use to which the word technology is made, and provides for the application of a single classification scheme to many kinds of organizations.

The empirical literature dealing with the incidence and general effects of "modern technology" in plant and office is based, at least indirectly, on one or another of the diverse classification schemes included in the collection above. This is the main reason why the results of the studies reported below may differ one from the other—different definitions of the situation produce different results. It is, however, not the only reason why the results will vary. There are at least two others. First, the results are not always comparable because of a tendency to classify definitively what might be considered in the interstices between traditional industrial techniques and automation—what may be more or less arbitrarily classified as sophisticated technology may in fact be more like old technology, for example. More extensive use of intermediate categories such as "in-between," "moderate" or "medium-sophisticated" technologies would be useful in this regard. The second additional issue in the comparability of results involves classification in terms of a machine or technique which, for some workers involved, has job demands of a more transitional nature. For example, a machine might be fully automatic, but transitional jobs are maintained, even though redundant because of a management concern regarding layoffs, a real need for these transitional roles during initial "debugging," or a management inability to forecast the necessary labor force in advance. In any event, jobs of this type are bound to be less fulfilling, less secure, and hence more disrupting and dissatisfying. As will be seen in the brief review below, to the degree we are able to account for the three contaminating factors just mentioned, the trends or constancies in organizational and individual behavior are strongly suggestive of the probable outcome of the emerging technology of the future.

The Evidence: Behavioral Effects

Little direct evidence is available for the proposition that fundamental differences in organizational form or type are related to technology. This controversy, centering in whether technologically advanced industrial organizations are becoming more or less bureaucratized or more or less autocratic, rages in general theoretical terms. Most available evidence used is segmental, its ambiguity attested to by its use by advocates on both sides.

In terms of general conclusions, Blauner (1964), Woodward (1965), and Thompson (1967), for example, all maintain that higher production technology is associated with more flexible and open organizations with more autonomous and satisfying work. Burns and Stalker (1961), Dubin (1965), and Harvey (1968), on the other hand, conclude that the organizational structure associated with modern production technology is more centralized, more rigid, with proliferation of specialized subunits and close supervision. The data provided for neither side seem irresistable and, as will be seen below, it is probable that the most generalizable statement lies somewhere in between (Anshen, 1962; Argyris, 1962). Most studies in the area are more fractionated, supporting or refuting only individual aspects of the above conclusions. From these, an empirically based theoretical mosaic can be constructed.

Blue-collar technology. In the area of blue-collar automation, as technology became more sophisticated, skill requirements increased in maintenance functions and not for production jobs as a whole (Faunce, 1958; Bright, 1958); and job responsibility or the requirement of staying alert and appropriately intervening in the event of trouble increased (Bright, 1958; Friedmann, 1961; Harbison, Kochling, Cassell and Ruebmann, 1955). Blue-collar responsibility also tends to increase with automation, with individuals supervising a larger part of the line and/or quality control (Bright, 1958). Static studies suggest, however, that supervisory (and perhaps employee) horizontal communication patterns can be functional at the lowest hierarchical levels in mass-production or traditional technologies (Faunce, 1958; Simpson, 1959; Jasinski,

1952). It also appears that at least so long as machine breakdowns constitute a problem, semi-automated or automated plants will require vertical communications patterns and perhaps closer supervision at the lowest levels (Harbison, et al., 1955; Mann and Hoffman, 1960; Woodward, 1965; Faunce, 1958; Dubin, 1965; Harvey, 1968). Walker's findings (1957), however, and those of Marrow, Bowers and Seashore (1967) reveal that, over time, the close supervision required to "debug" a system becomes less necessary and eventually decreases. These same authors, as well as Mann and Hoffman, provide data which show that although new technology may initially create constraints on supervisory behavior to closely supervise machine operations and breakdown, the behavioral effects of these constraints are not necessarily resented by the workers but are frequently welcomed. Close supervision required for the continuing nature of increased skill demands, on the other hand, may continue (Harbison, et al., 1955; Bright, 1958; Dubin, 1965). The outcome of such close supervision may actually be in the direction of greater work autonomy and flexibility, rather than less, with actual work operations simply being moved up one hierarchial level. As supervision becomes "closer and closer" and more and more highly skilled, the point is approached where the work organizations in automated factories may begin to resemble that of craft industries where the supervisor is simply the most skilled of the workmen. If the group continues to decrease in size, it may eventually be considered either a leaderless group or a "group of supervisors." The supervisor will become more a controller of output—not necessarily a supervisor of people, but of machines.

Technological effects on the blue-collar work group. The effects of modern technology on the nature of the blue-collar work group seem less equivocal than those on blue-color communication and skill requirements mentioned above. In two cases of unplanned social change following technological change (Mann and Hoffman, 1960; Walker, 1957), intragroup status differences were reduced and work roles became more interdependent under advanced technology. Evidence that this kind of effect on the blue-collar work group is a useful sort of target of planned social change is presented by a group of researchers at the Tavistock Institute in London (Trist and Bamforth, 1951; Rice, 1958, 1963; Trist, Higgin, Murray and Pollock, 1963; Thorsrud, 1968). The Tavistock people conclude that more sophisticated technology can lead to a number of antithetical supervisory and work group behaviors, but that autonomous group functioning (multiskilled workers, responsibility to allocate members to all roles, group incentive payment, and task definition involving continuity) seems to have the best results and is a more natural out-growth of the technological change itself. The Tavistock people make an important contribution by stating that: *More sophisticated technology is a necessary condition in instituting autonomous groups, but for best effect, that group structure must be consciously installed* (Trist, et al., 1963, p. 293). This aspect of technological effect is part of a rather thorough and extended theoretical discussion of socio-technical systems.

In discussing the autonomous group and socio-technical systems, the Tavistock people describe a different and broader role for the supervisor from what is traditionally held. It is required that, as the group comes to control the production process, the formal supervisor shifts to a control of boundary conditions such as maintenance and supply (Emery, 1959; Rice, 1963, p. 8). This is similar to observations of the emerging supervisory role by Walker (1957), Mann and Hoffman (1960), and Marrow, et al. (1967). The easier technology makes it for both the group and the supervisor to evaluate results, the easier it becomes to supervise on the basis of results and autocratic management of work activities is less likely. Even though the pattern of actual supervisory activities differs from one industry or plant to another, it is apparent that the trend with advanced blue-collar technology is in the direction of supervisors doing less in the way of traditional management—i.e., supervising the behaviors of others and attending to selection and training functions— and more in the direction of either acting as a facilitator and communications link for the work group or becoming more technically skilled operators themselves.

White-collar technology. Much of the evidence available for the behavioral effects of white-collar automation shows that they are not very similar to those of modernization of technology in factory work. Ida Hoos (1961), for example, in a study primarily of keypunch operators, concludes that unlike factory automation which has been said to enhance intrinsic interest in the job by integrating functions so that workers are not assigned to fragments of the production process, automation in the office has splintered job content into minute, highly repetitive units which can be processed by the computer. Faunce (1968, pp. 47-48) maintains that current development in Electronic Data Processing (EDP) is increasing the number of semiskilled operators. Office automation, he says, may better be called a major advance in mechanization of information processing. Only improving computer input-output systems will lead to true office automation. He likens pre-computer office work to the earlier craft or handicraft period in that the equipment (e.g., typewriters) do not have the skills built into them. Now that machines are being given these skills, the operators become semiskilled.

Other evidence (e.g., Mueller, 1969, Chapter VI) reveals that work in the plant and office is becoming more alike—jobs are more important, require more responsibility, and are more demanding. These jobs may differ, however, in that office jobs involve fewer duties. In one study which showed that white-collar clerical employees in the new work environment had more interesting and challenging jobs than before, the greater exposure to risk and tighter performance standards negated these attractive aspects in the content of the jobs. Centralization of control and decision-making followed the greater integration in the system. Autonomy and flexibility were reduced. Supervisory tasks ceased to exist and lead-clerks were eliminated. (Mann and Williams, 1962).

There are no published findings that technologically-induced job changes in white-collar work have much effect on supervision or work group structure. White-collar work, it appears, is a hybrid. Simple changes in computer-input form would drastically reduce the number of keypunch and set-up jobs. Since job responsibilities and availability of impersonal worker feedback have already increased, it seems likely that the character of white-collar work in EDP and jobs in automated factories is more similar than different if we exclude the great number of repetitive, fractionated, transitional tasks in white-collar work. These transitional keypunch tasks are of the type mentioned earlier, and are similar to those in factory automation, but are simply greater in proportion because of the present nature of the EDP "hardware."

Summary. It would seem that automation does in fact provide the potential or opportunity for enhanced worker discretion, responsibility, intra-work-group autonomy, interdependence and cooperation in both blue- and white-collar organizations. In fact, what seems to be happening is that lowest level jobs are becoming more like traditional supervisory roles. These behavioral effects seem very similar to the sorts of managerial prescriptions advanced by such management-organizational theorists as Likert (1967). These results are not unequivocal, however, especially in white-collar work, but this anomaly seems primarily a function of the somewhat lagging qualitative position of white-collar automation, and not of an intrinsic or quantitative difference.

The Evidence: Attitudinal Effects

Attitudes hold an interesting position in the causal matrix of behavioral change. Attitude can at once be considered a source of behavior and behavioral change—a predisposition to behave (Allport, 1954)—and an outcome of behavior as well (Festinger, 1957). We are using it here as the latter, a dependent variable. The previous section discussed the evidence for behavioral concomitants of technology. The present section describes the attitudinal resultants of this behavioral change. These attitudinal components in the studies reviewed here are, in the main, satisfactions. For our purposes, satisfactions are perfectly reasonable measures of the degree to which employee needs are fulfilled (Maslow, 1943; Morse, 1953). One must be careful, however, in attributing validity to even the most reliable satisfaction measure, in that satisfaction of one aspect (perhaps not measured) could conceivably manifest itself in another associated aspect or need area. This means that, at worst, we may say that technologically determined aspects of work are associated with enhanced feelings of satisfaction in general and, at best, that these aspects and behaviors lead to feelings of satisfaction in specific areas, hence representing propensities to maintain the conditions and continue the behaviors. The determination in each case must rest on the situation reported and the measures used.

The confounding effects of rural-urban values. Before considering the attitudinal concomitants of modern technology, several related methodological issues should be raised regarding the literature to be discussed. Turner and Lawrence (1965), studying the effects of simple and complex job technologies, discovered that their predictions (that the larger the job and more complex the technology, the better the work performance, attendance, turnover, and worker attitudes) held much better when they separated those workers who were from small towns from workers from large cities. Many more of the predictions held for the former and results tended to reverse for the latter. In simple technologies, where they expected unfavorable worker response to the job, they found that town workers, in fact, responded unfavorably, but that city workers tended not to. This latter group was concerned with external rewards and equity. They maintained low involvement and were unconcerned about changing things. In complex job technology, on the other hand, city workers had unfavorable responses, but the town workers enjoyed the intrinsic rewards and cherished especially the autonomy component (pp. 121-129). Further, these researchers found that the generally favorable attitude of city workers to simple tasks led them to distort their perceptions of these tasks and to say that they had considerably more opportunity to contribute than they did according to the researchers' ratings of these jobs. Town workers, on the other hand, perceived their work much as the researchers did (p. 87).

Hulin and Blood (1968), in a review of the literature on job enlargement, conclude that the case for job enlargement has been drastically overstated and overgeneralized. Enlarged jobs which motivate workers, decrease boredom and dissatisfaction, and increase attendance and productivity are valid only when applied to nonalienated (i.e., nonurban, and accepting middle-class norms) blue-collar workers, all white-collar workers, and supervisory personnel. In a study of the relationships of cultural differences (alienation from middle-class norms, or lack of it) and job enlargement, Blood and Hulin (1967) found that when white- and blue-collar workers were combined, no effect was manifest, but that analyzing blue-collar data alone revealed that the degree of urbanization and income were clearly negatively related to worker acceptance of larger jobs. Hulin and Blood (1968), however, speculate that being *placed* on a high-skill-level job may be qualitatively different from having a present job *enlarged*. They assume that correlational studies can provide evidence on effects of basically manipulative programs. Since what we have been discussing as concomitant with technological change can probably be classified as new, qualitatively different work, it follows that it may be different from ordinary job enlargement attempts where only the job itself is changed. In any event, Hulin and Blood provide a possible reason for some of the incongruities found in the literature regarding attitudes and technology.

Examples of possible effects are numerous. Mann and Hoffman described the modern plant they studied as being staffed by a combination of volunteer workers from the old metropolitan plants, and young inexperienced workers from the rural area in which the new plant was located. They found also that workers in the new plant more frequently expressed a "Republican" political orientation than workers in the old plants (1960, p. 40). The data these researchers collected may well reveal the effects of newly installed technology on nonalienated blue-collar workers rather than blue-collar workers in general. Several other studies, although not specifying the urbanity of the plant involved or workers' political attitudes, did indicate that the automated installation was manned by volunteers (Faunce, 1958; Rice, 1958; Walker, 1957) which suggests that in so doing these workers revealed an openness to change more characteristic of the "nonalienated" as described by Hulin and Blood.

Blue-collar attitudes and technology. Evidence for satisfaction in blue-collar automated work is considerable (Mann and Hoffman, 1960; Rice, 1963; Thorsrud, 1968; Blauner, 1964; Woodward, 1965; Walker, 1957). Although both Faunce (1958) and Walker (1957) found that satisfaction with new technologically created work group organization and environment was initially low, Walker in his four-year study found that many changes took place in worker attitudes. After three years in the new mill, Walker found that workers were considerably happier (pp. 73-100). They were pleased with the group incentive system although they wanted credit for down-time and the inclusion of maintenance people. Interaction with supervision had decreased over the previous period, but the strong worker hostility of the year before was gone. Because the new mill had been "debugged" and was operating well and interest in production was up because of earnings, the supervisors no longer pressured for keeping the machines running. Teamwork had increased; workers were using the new public address system; they needed little technical help and they felt "at home" with the machine; they had become more autonomous. Jobs were termed "mentally harder, but physically easier." The few nonautomated jobs in the mill, at first popular, became resented. Newly-won skills and a lower fear of the machine led to greater job satisfaction; the ability to occasionally override some of the automatic features in order to produce greater output improved morale still more. Workers still felt management was not supportive in reducing the amount of smoke in the shop, but more important, according to Walker, management had not acted quickly enough or informed the workers as to what they were doing about it. Walker reports that the greatest satisfaction unverbalized by the workers was the satisfaction produced by the rhythmic pattern of interdependent motions performed by the group (pp. 104-105). In less favorable attitude areas, workers were even more negative in their feelings about promotion than they had been the year before. They resented the influx of technical, educated management. In total, after three years, the workers' attitudes were very favorable regarding the job and the work, less favorable toward management (because of the smoke), and quite negative about the displacement effects and lowered promotion changes.

Woodward (1965) presents cases of technological change in automatic packaging and an automatic assembly line. The dissatisfactions created initially were great—nothing suggested that these plants would elicit the same feelings as those in the other technologically advanced firms which she studied. During the study, however, workers did become satisfied with the work. Woodward presents this as evidence for Festinger's thesis that long-term satisfaction can follow real dissatisfaction at the time of change (p. 236).

Marrow, Bowers and Seashore (1967), and Seashore and Bowers (1970), in their long-term study of technological-structural change, present results which support Walker's finding that workers' attitudes took longer to change than did their behaviors. These results imply that the slow and small change in attitudes they initially measured, relative to changes in perceived supervision and increased productivity, indicate that more time than the original study covered was required for attitudes to change. This may be especially true in this case, where the organization under study was evidently in a very poor condition at the time the study began.

It seems clear that factory technology affects worker attitudes through the choice of work methods and social system employed, at least when measurement of those attitudes follows technological change by some time. What is not clear is whether positive effects can be produced for all workers, and not just the nonalienated. Mann and Hoffman demonstrated that rural people and volunteers can react favorably to a technologically-determined work situation where people are given credit for being more than adjuncts to machines. Walker's study suggests that, over time, an unplanned but satisfying social system can emerge which is strikingly like the Tavistock "autonomous groups" system, and which is a direct outcome of a technologically-determined work situation. Walker's new mill, however, is in Lorain, Ohio, a small town; the original work group was composed of volunteers. Faunce, on the other hand, did not find positive effects, but his study involved a static design, undertaken shortly after the new equipment was installed. In this sense, Faunce's results and Walker's results after nine months are remarkably similar. Blauner's results, although tending to suggest that technological effects over-ride the cultural ones, are not supported by direct attitudinal data; nor does he attempt random sampling (e.g., for town-city) within the industries he studies. Finally, Woodward presents data which suggest that the phenomenon of workers' eventual satisfaction with a change situation comes about by perhaps simply "getting used to it," or what Friedmann has called "habituation" (1961, p. 8).

White-collar attitudes and technology. The empirical effects of the new, computer-determined work on attitudes of white-collar workers is equivocal. Some positive effects are noted, (e.g., Mueller, 1969), but negative effects dominate the literature. Mann and Williams' study of white-collar automation (1962) reveals that in a five-year period following installation of the computer, a significant change in satisfaction with various job aspects took place. Employees in the departments affected became more satisfied with the increased responsibility, increased variety, increased opportunity to develop and learn new skills, and they saw their jobs as more important. Employees in unchanged, control departments reported no such changes in their satisfaction. Neither group, however, significantly changed its overall satisfaction with the job. The researchers interpret the findings as indicating that the increased demands on the changed department for more and better work negated the positive effects of job enlargement on job satisfaction. They also reported that a significantly greater proportion of employees in the jobs affected indicated their worry about security in the later survey. They conclude that in the early stages of technical changes, supervisory technical skills are most important, but as the system is debugged and running smoothly, supervisory human relations skills become more important (1959).

A study of automation in a French bank (Marenco, 1965), which utilized measures over time and a comparison of changed and unchanged departments, revealed negative attitudes. Attitudes before the change were disparaging. Most respondents felt that their jobs were uninteresting. Sixty percent were satisfied with the work and had little desire to express initiative and no desire for responsibility. Seventy percent had indifference to anything but their immediate job. At the time of the change-over, employees in the affected departments were asked to participate in decision-making and were told they were elite. These respondents in the second survey thought their jobs were the most important in the bank. Programmers had higher job satisfaction, but the keypunchers' satisfaction with the job was no higher than their unchanged clerical counterparts. All respondents in the departments affected were more satisfied with the company and work atmosphere. Most thought that management expected more initiative and suggestions from them. They said that they found adapting to the new work difficult and that they expected more supervision and more impersonal relations with their supervisors. By the third survey, job satisfaction in the departments affected was down to the level of the bank as a whole. Respondents in these departments still felt that their jobs were important, but no longer felt a desire to exercise initiative. Clerks in these departments affected wanted to be strictly supervised, and programmers wanted to be left alone. In general, the attitude was one of indifference. These results are strikingly like those of Walker described above, at least like his initial findings. It would appear that, for Walker's steel mill, management provided the technical help necessary to get the operation running smoothly, then pulled away and allowed workers to operate the system more autonomously. In Marenco's bank, on the other hand (without any additional evidence), it seems that management failed to help get the system working smoothly; employees felt uncomfortable in their lack of success and ultimately demanded less autonomy rather than more. It appears also that supervisory style in the bank at the time of changeover had a human relations orientation rather than a technical orientation which Mann and Williams propose as more efficient initially.

Mann and Williams' finding regarding technologically-determined job enlargement in white-collar automation, and Marenco's findings regarding motivation and initial desire to participate in technologically-stimulated work decisions are suggestive of the possible enhancing effects of computer installations on clerical workers, but the findings of each are obscured by what seems to be inappropriate management in both cases. In the Mann and Williams study, management's technical emphasis may have been so strong, and pressure for production and quality so great early in the changeover, that people in departments affected were alienated. In Marenco's study just the reverse seems to have taken place with management emphasizing human relations skills when it should have been more technically oriented and attempted to help employees become competent in their new jobs.

Summary. The behavioral effects of automation seem to produce favorable attitudes which may in turn maintain those new behaviors, especially when the effects of time following the behavioral changes and the effects of alienation from middle-class norms are accounted for. Although this evidence was obtained primarily in studies of blue-collar organizations, the limited data in white-collar organizations suggest that the same effects can be achieved there as well. These effects, in any case, are not directly resultant from behavioral constraints of modern technology, but obviously are conditioned by management decisions to facilitate living with the new technology. Two cases of total attenuation of favorable affect following behavioral changes as a function of management decisions were described.

Strategies for Organizational Change

Internal Subsystem Interdependence and Strain

The evidence cited above suggests that technology can affect organizational structure (e.g., Woodward; Harvey; Burns and Stalker; Blauner), behavior, and productivity (e.g., Trist, et al.; Rice; Walker; Mann and Hoffman;

Mann and Williams; Marrow, et al.), and attitudes as well (e.g., Mann and Hoffman; Walker). It seems clear that there is a systemic interdependence among the subsystems of an organization. Changes cannot be effected in the technical system without repercussions in the social system. Katz and Kahn's description of open system theory in organizations (1966, pp. 19-29) enables them to deal with the relatedness of subunits or parts of a system vis-a-vis the organization's environment. These authors see a relationship between the necessary effect of the interrelatedness of subsystems and the degree of organizational change which can be effected. Parts of social systems, for example personal and role relationships within work groups, can be changed without manifest effects on technical systems or on organizational structure, but the change in organizational behavior is mild and bland reform, not radical change (p. 424). By the same token, parts of technical systems can also be changed without undue stress on the other systems in the organization. Woodward found, for example, that technical change which did not radically affect the nature of the production system only resulted in minor modifications to the organization. She claims that greatest changes will be noted when a shift is made from large-batch to continuous-flow production (1965, p. 72). Katz and Kahn view systemic change as the most powerful approach to changing organizations. Systemic change involves changed inputs from the environment which create internal strain and imbalance among system subunits. It is this internal strain which is the potent cause of the adaptation of subsystems indirectly connected with the change input (pp. 446-448). Katz and Kahn state:

> The basic hypothesis is that organizations and other social structures are open systems which attain stability through their authority structures, reward mechanisms, and value systems, and which are changed primarily from without by means of some significant change in input (pp. 448-449).

Values and motivations of organizational members change, Katz and Kahn (p. 446) maintain, in a more evolutionary way and are not as immediately amenable to the influence of changed inputs as is organizational behavior.

Guest (1963, p. 55), Mann and Hoffman (1960, p. 193), Marrow, et al. (1967, p. 229), and Woodward (1965, p. 239), among others, support the notion of interrelatedness of subsystems and the importance of considering the derivitive effect on the social system of significant changes in technology. They all conclude from their evidence that a sound technical system was not sufficient in itself to assure good performance via the existing social system.

Order and Precedence of Subsystem Change

It seems established that when technological change is considerable, some effects on the social system must be recognized and planned for, but the question of coordination of change in these two systems is still unanswered. Is technical change the best way of achieving organizational change, or would it be more efficient to purposefully change the social system, following that by planned changes in the organizational technology, or to change both simultaneously?

Cultural anthropologists have long maintained that changes in technology historically lead to changes in attitudes, values, and philosophies (Mead, 1956; Ogburn, 1957, 1962; White, 1959, p. 27). Some direct evidence for this position is found in the studies reviewed here. Burns and Stalker, for example, concluded that mechanistic and organic management systems were dependent variables to the rate of environmental change (i.e., technology and market situation). Trist, et al. (1963, p. 293) state that change conditions for installing autonomous groups were more favorable under greater mechanization and low group cohesiveness. Even here, however, these authors discovered that when new equipment was installed, the existing social system could create forces of resistance to the full potential of the technology. The technical outsiders supervising the change were frequently unaware of the operators' responsibilities to the rest of the work cycle, and tended to isolate the machine activity (p. 273). This had effects of unfavorably disrupting the existing social system and making subsequent change difficult. Planning for this contingency was necessary. Guest, (1962, pp. 52-53), and Marrow, et al. (1967, p. 237) attribute direct behavioral effects to the relatively minor technological change made, but, in both of these cases, these changes were preceded by management succession. These can be considered cases of what Schon (1967) describes as "innovation by invasion." The old borrows what it can of the new; the new introduces change into the old, or the new displaces the old. In the open system notion of Katz and Kahn, this is systemic change representing organizational social system change via new inputs from the environment. In both of these cases, changes in interpersonal relations were modified at the top of the organization coincident with the technological change at the bottom.

Precedence and hierarchical level. Argyris (1962), like Anshen (1962), suggests that production technology has little effect on higher management. Argyris continues that effective organizational change comes about by improving interpersonal competence directly at the top of organization, while improving it at the bottom more indirectly through changes in technology and control systems (p. 282). This seems similar to the Tavistock notion that the socio-technical system operates primarily at the lower part of the organizational hierarchy, unlike the total organization dynamic of Katz and Kahn's open system. The Tavistock group, although it makes excursions into the socio-technical nature of management systems (Rice, 1963; Trist, et al., 1963) is primarily concerned with the production system (e.g., crews of miners across shifts) as socio-technical systems. It is implicitly clear in these particular studies, however, that effective introduction of technological change for ultimate organizational change involved the upper ranks, either in a commitment to plan adequately for social system effects (Trist, et al.), or in a commitment to the technological change itself as a method of improving social relationships (Guest; Marrow, et al.).

Precedence and organizational requirements and resources. The cases studied by both Guest and Marrow, et al. involved organizations faced with a need to adapt quickly to a threat to their existence. Previous structure and technology were incapable of eliciting the motives, the skills, and coordinative efforts demanded by the environment. In the case of Marrow, et al., technical change came first, followed by the planned change of the social system. This priority was taken because management had the technical skills but fewer resources for dealing with the human organization (1967, p. 21). Guest's study involved changing the social system first, then changing the technical system, and then allowing technological effects on social system to continue and strengthen. The new top manager in this case was able to institute group processes in problem-solving throughout the organization because of his own personal ability to elicit trust and confidence in all persons in the factory (e.g., walking around in shirt sleeves, requesting rather than commanding, etc.). Another manager with less "natural" or comfortable interpersonal skills may well not have been able to do this. It seems reasonable, therefore, to speculate that technological change probably best precedes the planned and anticipated social system changes at the bottom of the organization.

In an example of similar successful social system change in a white-collar organization (O'Connell, 1968), the outside consultant focused on the company as an ineffective technical system rather than as an ineffective social system and set out to maximize the efficiency of that technical system within the constraint of the existing methods of management (p. 124). Although the consultant used change methods of management training and exhortation for changed supervisory behavior (direct attempts at social system change), he also restructured the organizational technology (in this case, patterns of relationships and work flow) so that lower managers were forced or constrained to behave in new ways.

Necessary Prior Conditions for Technological Change

Even though it is apparent that the theoretical models of open systems and socio-technical systems underlying organizational change strategies are relatively new and untested, there has been a good deal of atheoretical writing about how management should institute technological change. Although we have already considered the idiosyncratic personality effects in Guest's study which allowed social system changes to precede technical change, the notion persists that to undertake successful technological change (i.e., little resistance and much social change), the company must first have good labor relations, high employee satisfaction with the company, and mutual trust and goodwill (Mann and Hoffman, 1960, pp. 199-200; Nangle, 1961; Scott, 1965, p. 93). This implies that if these conditions are not met, then management must undertake to improve them before considering a technological change. There is without doubt some absolute ceiling on the amount of negative feelings and poor labor relations below which successful technological change should not be expected, but it will be maintained here that such a ceiling is quite low. It may be in fact that when conditions are below this ceiling, alternatives open to the organization are limited to selling out, wholesale dismissal of disgruntled members of the work force, or moving to a new location to start afresh. Certain more limited kinds of necessary prior conditions with higher ceilings, however, probably do exist. Workers' sense of job security in face of technological change would be an example. This condition can be met by meeting certain minimum guarantees. These guarantees include consideration of stable wages, shorter hours of work, unemployment insurance or guaranteed annual wages, and retraining programs. This list perforce implies some absolute minimum openness in the organization, but emphasizes economics-based labor relations.

Another stated management approach to technological change is the "sales job." Bright, for example, claims that it is probably better to go slow in introducing automatic machinery and to sell and familiarize the workers on the advantages and outcomes to insure their acceptance (1958, pp. 210-211). He provides an example, however, where the management persuasion attempt emphasized advantages, but neglected mention of the need for fewer workers. Successful change was achieved as workers ". . . became familiar with the new plant, [and] as the troublemakers were eliminated . . ." (1958, p. 198). This is clearly an example of organizational conditions of openness being so low that social system change is required before technological change could be effected. The sales job here seems irrelevant. Nangle (1961) found that the degree to which the company pre-change information program or sales job was believed by employees was related to a favorable shift in their scores on a Readiness to Change Scale. Nangle's data are from a study where measurement was made before and after a changeover to EDP. The shift in Readiness to Change followed the actual change as a function of employee trust in management communication. Absolute levels of belief in management communication, if low, would suggest that subsequent resistance to change could be reduced by exposing employees to a number of prior changes.

In line with the finding by Nangle, Mann and Williams (1959) state that technological change will be easier in a company with a history of change every few years because employees are more likely to have learned to accommodate to such changes. Marrow and his colleagues point out, however, that although such a company will be easier to change, it will probably be less in need of change than a company characterized by rigidity and stability. Marrow, et al. also state that a company with an antiquated social system will probably also have an antiquated technology (1967, p. 240). The apparel factory they worked with was in dire need of change but had neither a history of flexibility nor much trust and openness. The change strategy they undertook assumed that one cannot ordinarily wait for the slow process of self-generated social change. Coercive steps by new management made success possible by coupling moral support and practical aid with the always pre-emptory and compelling demands (p. 240). Schachter, Willerman, Festinger, and Hyman (1961), in a study of job-related change, discovered that people working at highly patterned jobs will experience more disruption in a job change if they are harassed prior to the change. Schachter, et al., after matching work groups for task, size, and productivity, treated them in different ways prior to a work change. One group was treated especially nicely, the other was harassed; there were no control groups. They found that close supervision after the change overcame some of the disruption in the negative groups. Instead of concluding, as they should have from these data, that close supervisory help may be especially useful during periods of change when attitudes are negative, they concluded that the organization and its members must be brought to a state of readiness for change before change is effected. Since no control data are available, it is speculative that the favored groups differed significantly from groups with normal (nonmanipulated) attitudes in their response to work changes.

Chapple and Sayles (1961, p. 88), in discussing results of an earlier study, conclude that acceptance of technological change is most favorable when work groups are either apathetic or erratic in grievance pattern. These patterns are a function of low job interdependence, geographic isolation, and the feeling that one is dispensable—that is, when little organized resistance is available through group cohesiveness. Trist, et al. support this in their finding that acceptance of autonomous group functioning following technical change varies inversely with the degree of local tradition which is unfavorable to the new social system (1963, p. 293).

What appears minimally necessary for successful technological change is not organizational sweetness and light, but a circumstance where management can guarantee minimal job and pay security, and where the labor force is not automatically, actively, and cohesively opposed to any management action. It does not follow from this that management should take an obscurantist attitude or that employee participation is not useful. It does follow, however, that measuring the potential resistance to change is much more important when people are told they must change, or when there is little or no sense of participation. These elements may not be necessary if it can be assumed that resistance will not be problematic.

Implications of Sophisticated Technology Effects on Social System Change

Direct effects. The direct effects of modern technology on more satisfying and productive methods of working have been discussed above (Guest; Mann and Hoffman; Mann and Williams; Marrow, et al.; Trist and Bamforth; Turner and Lawrence; Walker). Although the evidence is not overwhelming, it appears that a properly planned technical change can lead directly and without additional inputs to increased job complexity, increased work complexity, increased work group processes, more helpful supervision, and higher productivity. It was also found that, at least for some workers, these direct effects can lead to more positive job involvement, improved work group relations, more favorable attitudes toward supervision, and pride in higher productivity. The implication here is that social change can proceed as a direct outcome of certain technological changes, but if this were the only way social change was effected, the outcome would be limited by chance factors.

Indirect effects. Several of the studies reviewed here consider social change not only resulting from the new technology itself, but resulting from a planned social change input made possible, at least in part, by the mere disruption created by the technological change (Marrow, et al.; Rice, Trist, et al.; Williams and Williams). This kind of disruption has been labeled "unfreezing" by Lewin (1951) or "internal system strain" by Katz and Kahn. In both cases, the dynamic created is that of a force toward total system restructuring to find a new equilibrium. It seems possible that social system changes are possible without technological change, but organizations may not be able in themselves to provide the force necessary. Williams and Williams (1964), for example, state that such changes are not possible without a catalyst-like expenditure on technological change which creates stresses forcing departments and units to compromise on objectives and abandon traditional routines and activities. Marrow, et al., claim that their apparel factory was unable to unfreeze itself; heroic measures were needed to create enough disturbance to allow normal change processes to begin (1967, p. 232). Trist, et al. maintain that even limited technological changes can create enough disruption if their potentiality for inducing social change is recognized (1963, p. 284).

The direct effect of technology on social system changes seems to involve the dynamic of constraints applied on employee behavior. The dynamic of unfreezing, on the other hand, seems to be a freedom provided by the new technology to seek new ways of behaving. Internal system strain comes about in the latter case in that employees' cognitive maps may not match managements' under these circumstances, or the values of one department may not match those of another. Management awareness of and planning for such conditions become important for the outcomes on the social system.

It seems clear that the combination of direct and indirect effects of technology on the social system provides the basis for concluding that technological change would best precede social change in that it probably requires less time and elicits less resistance. This is true because technology not only disrupts or unfreezes, but imposes strict, nonhuman controls on minimum behavior.

Marrow, et al. cite four general approaches to changing individual behavior: (a) altering the working environment, (b) altering individual percepts and cognitions, (c) altering individual motivation, and (d) implanting new skills (1967, p. 234). It has been implicit in the above that technological change leads directly to social system change via (a) and indirectly to social system change via (b), (c), and (d). It has also been suggested that of the indirect forces (d), skill training, is surely necessary for effects on both behavior and attitudes; (c), motivation, is necessary for predicting effects on attitudes; and (b), exhortation, may not be necessary at all, but simply desirable. It is suggested here that because of this pattern, and the fact that direct social system change involves only exhortation, resistance to social system change will be lower following technological change than preceding it, and that this resistance, following technological change, is attenuated as a function of the amount of situational constraint applied by the technology. It is also maintained that technological change preceding social change will meet with less resistance than social change undertaken before technological change.

In recent books dealing with general effects of organizational systems, Buckley (1967), and Thompson (1967) suggest that technology can provide clear, rigid situational constraints on behavior, while normative rules (e.g., role structure) have a wider range of permissible behaviors. Resistance is more easily measured in the former case because it is easier to note deviance from clear rigid rules than it is to tell deviance within the confines of a range of behaviors. Buckley, for example, says:

> Norms and values, and hence role and institutional structures in general, do not specify concrete behaviors; they are more or less general rules or guides and do not contain enough information to specify the detailed operation of the system . . . (1967, p. 159).

Roles, as the traditional agents of structural influence on the individual, may produce desired behavioral changes but they must be integrated with the structure and not treated in isolation. Katz and Kahn admit that the cumulative effects of small, internal organizational changes in the social system (e.g., roles) may in time produce organizational transformation of great depth without the advent of external structural forces, but the lack of evidence for this makes it speculative (1966, p. 449). Social change involved in technological change is more difficult to resist for it would require sabotage of the equipment or process which would be immediately obvious. Documented reports of recent "Luddite-like" activities are not numerous and, when found, deal exclusively with resistance at middle-management levels (American Foundation on

Automation and Employment, 1966, pp. 28-29; McKinsey and Co., 1958, p. 6).

Situational constraints can include roles, if roles are clearly integrated with the technology. Thompson and Bates (1957) state that if the ratio of mechanization to professionalism of an organization is high, then an engineering approach to changing behavior is required. If the ratio is low, then the coordination and integration of human activities through the role structure is the major concern. O'Connell's case study of major organizational change (1968) treated the role structure of the insurance company as the system of technology. In this case, the target of change was the behavior of first-line sales management. Change was accomplished by planning detailed modifications in the structural context and role set surrounding the manager. Virtually all of the formal design was engineered not as an end in itself, but as a means of effecting a behavioral change in sales management (p. 68). Role senders (superior, subordinate, and at peer levels as well) were all told of new expectations for the sales managers. Given these systemic alterations, a sales manager could not "comfortably" deviate very far from the proposed behavior pattern (p. 136). Since the technology in an insurance company is more a process than a machine, these situational constraints, although clearly role changes, are much more engineered than the usual method of changing roles by appealing directly to the change target to modify his behavior. O'Connell compares the environment, structure, and behavior itself as potential sources in changing behavior, and concludes that although structure and direct appeals for behavior change are of equal importance and directness in changing behavior, structure is more controllable in effecting behavioral change (p. 56).

Summary. Few direct, controlled results are available in support of, or in opposition to, the notions put forward by Katz and Kahn regarding notions of subsystem interdependence, and efficient precedence of structural and technological effects prior to social system change. Even with this paucity of evidence, the ideas themselves seem reasonable, and available data are at least partially supportive of them. Thus, it seems reasonable to postulate that technological change efficiently precedes attempts at planned social system change.

Technology, Role Constraints, and Permanence of Social Change

There is no evidence to prove that any change strategy provides permanent effects. (Schien, 1961), in describing change by "compliance" (where the individual behavior changes because the situation forces him to change), suggests that coercion-compliance is only a method of changing behavior; attitudinal change need not follow. In fact, if acquisition of new attitudes is also via coercive-compliance, these attitudes will be very temporary if they obtain at all. On the other hand, he continues, if behavioral changes are coerced at the same time as unfreezing operations are undertaken, actual influence can be facilitated if the individual finds himself having to learn new attitudes to

justify the kinds of behavior he has been forced to exhibit. These new attitudes should act then to maintain the new behaviors. This is exactly the
outcome that Festinger's dissonance theory of attitude change would predict
(Cohen, 1964, p. 82; Festinger, 1957, pp. 94-95; Insko, 1967, p. 219) that
counter-attitudinal role playing will result in consistency-producing attitude
change and maintenance of coerced behaviors. Although the attitude change
portion of this position has received consistent support in the psychological
laboratory (Insko, pp. 219-223), there is no mention of it as an explanatory
concept in field research around technological change, and no evidence for the
predictions of behavioral maintenance. In fact, only two cases exist where
attitudes in industrial work were found to change as a function of the relatively ambiguous condition of role change (Lieberman, 1956; Tannenbaum,
1957). As this coercion-compliance position has received little empirical support in the field, so has the other generic strategy—that of conversion (Bennis,
1966, pp. 170-171). "Conversion," as the attempt by persuasion and influence
to change the individual's cognitive or attitudinal set, is a more common
strategy in affecting social system change in organizations (Sayles, 1962).

Thus, it appears that although the evidence exists for asserting that
planning for social change should, where possible, take place around technological change, it is not unambiguous. It seems that when technological
changes are undertaken first, total resistance to changes will be lower (given
certain basic conditions are met), and the noticeable effects of changed behavior will be manifest earlier. It is not at all clear, however, whether this sort
of change strategy will work so well with people with certain characteristics
(for example, the alienated), or will have more permanent effects via changed
attitudes and satisfactions acting to maintain new behaviors.

It is not advanced here that the social system in an organization cannot
be changed directly by appeals to members to change the way they believe
and behave, coupled with attempts to train members in the skills necessary to
behave differently. In fact, it is assumed that such direct changes can be
effected. An interesting question to ask, however, is what facilitating effects
are manifest where the members affected by direct social change attempts
exist in a modern system of production technology—i.e.: (a) where behavioral
constraints may exist in the direction of greater worker discretion, responsibility, interdependence, and cooperation; (b) where some residual effects of
"unfreezing" may still exist in the direction of search for new ways of behaving vis-a-vis the modern technology; and (c) where the permanence of the
technological constraints on behavior and the unfreezing effects may combine
to change attitudes as well as behavior. Such facilitating effects seem reasonable to hypothesize given the data reported in the literature described above.
The question is not whether direct social change attempts can be effective,
but rather, whether the existence of modern, sophisticated technology can
enhance the results of such change efforts.

Conclusions and Implications

The Tavistock researchers have stated very clearly that sophisticated production technology, via the dynamics of socio-technical systems, makes change
in the direction of autonomous groups (or groups with high responsibility,
high participation in decision-making, and high self-leadership capability) much
easier. The Tavistock literature provides, without doubt, the best developed
theory of technological and social system interaction. Although the ideas are
provocative, the evidence used in supporting this viewpoint is meager—a small
number of intensive case studies in Great Britain, Norway, and India. Because
case studies were used (as they rightfully should have been, being exploratory), no attempts were made to assess technological alternatives, or to utilize
quantitative methods. Other studies dealing with technology and worker behavior do not bear on the Tavistock thesis directly. Those done around technological change confirm the Tavistock notion that sophisticated technology
can elicit behaviors consistent with those of the autonomous group. Still other
studies suggest that the existence of sophisticated technology (static technology, sometime after change), is associated with behaviors consistent with
those of the autonomous group concept. The results of these studies reveal
that technological effects are stronger in studies done in blue-collar than in
white-collar organizations. Since Tavistock data do not include white-collar
organizations, this *may* mean that the notion regarding facilitation of social
change is less useful in the white-collar case. Most all the studies use rather
unquantifiable, or otherwise limited methods of assessing technological sophistication, so none can be replicated directly or be considered directly comparable with the others. No studies done outside the Tavistock framework,
however, (aside from the study reported by Marrow, et al., and possibly that
reported by O'Connell), directly test the idea of sophisticated technology
facilitating planned social change. These other studies, however, tend to confirm the Tavistock socio-technical assumption that modern technology has
general and inherent forces in the direction of more democratic management,
and that this can be exploited in attempts to institute planned social change.
We have also noted that because direct effects of changed behaviors on attitudes are evidenced in the technological change literature, the changes effected
via the socio-technical system may not only be faster, but may be more
permanent (through effects on attitude change, and subsequent reinforcing
effects of attitudes on behaviors) than social system changes effected in other
ways.

The present study is an attempt to test these ideas more directly using
large-scale, quantitative survey methodology in longitudinal application. An
attempt will be made to construct a more replicable method of assessing
technological sophistication, and to use this assessment as a control in examining the results of planned social change programs in a white-collar and a

blue-collar organization. Based on the generally confirming results discussed in this chapter, the idea of technology facilitating social change, derived from the socio-technical theory of production system dynamics, seems not only provocative, but plausible as well.

II

THE MODEL AND HYPOTHESES

The conclusions drawn and questions asked in Chapter I form the nexus of the present research. The study presented here is an attempt to answer the question: To what degree does the sophistication of work group production technology facilitate attempts at planned social change in industrial organizations, specifically, change in the direction of work group autonomy and self-leadership? Facilitation as used here refers both to degree of change and implications for permanence of change as well.

This research will attempt controlled measurement over time of the success of planned social change strategies in organizational subunits using production technologies representing different levels of sophistication. The study will include: (a) a broad-gauge definition of technology, one including definitional components found useful in prior research, (b) a sample which includes white-collar workers (both technical and clerical), as well as members of manufacturing organizations, (c) some measure representing degree of social alienation from middle-class norms, and (d) measures of changes in attitude as well as changes in behaviors.

The major aim of this study is to bring to bear the accumulated knowledge described above in a test of the general conclusions derived. As the paragraph above suggested, one of the critical factors in such a study would be a more general taxonomy of industrial production technology. The task at hand then becomes the development of such a new classification model, seeing how well it fits the world of tasks and organizations, and how useful the predictions from it will be.

A Model of Technological Classification

The following is an attempt to build a taxonomy of direct *production* technology, and it is limited to that reference. *Direct production technology is here defined as the principles and techniques used to bring about change toward desired ends in the raw materials processed by a job or work group.* The job or task is a central concept here and refers to that portion of the employee's work role which deals with activities directly relevant to the creation of the product; it does *not* refer to the organizationally relevant, but nontask related interactions of people. This latter point, the exclusion of the social system from the technological classification, implies, for example, that managerial philosophy and style (although operating to change or control the behavior of employees, and hence a technology of sorts) has no *direct* bearing on the creation of the product. With supervisory and managerial activities excluded, it becomes clear that the reference used here is that of the task-oriented behaviors of nonmanagerial or nonsupervisory jobs, be these the jobs at the lowest, or production-worker level of the organization, or jobs involving higher level, professional, or technically specialized staff personnel. The supervision or administration of the production technology will be considered conceptually separate and are not expected to fit neatly into any taxonomic ruberic enumerated below.

As discussed in Chapter I, extant classification schemes vary widely one from another. In an attempt to build a taxonomy which may be used to classify technologies in the service industry as well as manufacturing, elements of the schemes already in use have been combined into a new model. This model incorporates (a) the nature of the raw materials, or production system input, (b) the nature of the "hardware," a dimension of production system throughput, and (c) the nature of the information exchange and evaluation as relevant to the production system output.

Raw materials. The first element, the input, involves a continuum of the operator differentiation of raw material. This continuum can be stated as ranging from low job-relevant sophistication of raw material—that is, little understanding or knowledge of, and much uncertainty about the material (which together provide only a limited definition and/or limited control of the parameters of the raw material)—to high sophistication—considerable understanding or knowledge of, and little uncertainty about the material. This sophistication or its lack lies *not* necessarily in the "essence" of the material, but in the perception defined for the typical role occupant. This perception as here defined has little to do with individual differences among role occupants, but is rather a function of organizational role definition and expectation, organizational selection procedures, and the objective state of scientific knowledge regarding the material. High sophistication in raw material can allow the worker or operator greater discretionary power and authority over the production options available. This increased discretionary influence and authority can be associated with increased responsibility for results. High sophistication could also free the worker from constant attention to the production process due to unpredictability in the material. The raw material "Corfam" in the shoemaking industry provides an example of both increased discretionary power and/or decreased need for attentiveness. A machine operator who is making shoes from the highly uniform synthetic Corfam and who knows something about the specific properties of the material would have considerable discretionary power and attendant responsibility in regulating the speed and feed of the machine within the limits of the material. This discretionary ability is significantly greater than that of the same operator making shoes of leather, which is a highly variable material with individual idiosyncracies usually beyond the knowledge of anyone. It is also true that the operator using Corfam could predict when and if problems would occur to a greater extent than if he used leather. By this token, he would not be as "tied to his machine" when using the synthetic as when using the natural material.

Equipment. The second element in the present model of technological classification, the throughput, involves a continuum of complexity of equipment. This continuum ranges from low sophistication—much human physical power and manual control—to high sophistication—much inanimate power and automatic control of routine operations. High complexity in this dimension provides the operator potential freedom from routine and monotonous tasks and hence the opportunity to undertake other more varied tasks. These could include supervision of a larger portion of the production process, coordination with others in the production process, and undertaking quality control and/or maintenance functions. More complex hardware can also imply increased operator responsibility for the more costly equipment.

Feedback loops. The third element, evaluation of output, involves a continuum of amount, speed, and sophistication of information exchange and evaluation. This continuum ranges from low sophistication in feedback characterized by simple evaluation—a "Go-No Go" message following the production manipulation by some period of time—to high sophistication in feedback characterized by a production control through a multitude of fast, broad-bandwidth channels carrying specific evaluative messages to the operator during the production process. Low sophistication or complexity, it should be noted, is a stimulus for correcting action *ex post facto*, or simple evaluation. High complexity, on the other hand, involves the regulation of current performance or control, as well as evaluation. Interest here is in the transmitted code, not the received code. The expectation of feedback redundancy, rather than the initiation of feedback verification refers here to the system rather than to the occupant and his abilities. High sophistication in this dimension provides the operator the opportunity to autonomously control his own behavior vis-a-vis the output process. The information received allows the operator to raise the quality of his production, and to produce more numerous successes.

Relationships among the three dimensions. The three technological dimensions described above have each been presented with examples of how higher technology in that case can lead to changes in the way work is done. They have, however, been presented alone, so that the job effects of one dimension or factor on the job effects of another are not described. It is assumed, however, that these job-related effects of increased sophistication in all three dimensions are compatible. It is also assumed that this condition *can* lead to the seeming paradox of "job simplification," *and* "role enlargement." It should be clear that what is described as a more sophisticated technology would actually result in a much smaller and simpler job. Job is defined as that *portion* of the employee's work role which deals with his activities *directly* relevant to his creation of the product. As technology has been defined earlier (i.e., a set of principles and techniques useful to bring about change toward desired ends), it involves the object to be changed and the object changing it. Man is always involved in technology as an extension of the machine or the object performing the change, or, as was clearer in earlier times, his tools are extensions of himself.[1] To the extent that man uses tools or techniques to effect change in some third object, three technological relationships can be expressed: Machine-product, Man-product, and Man-machine. The first two of these three relationships completely define the production process. The first, Machine-product, defines that portion of the process under automatic control. The second, Man-product, defines that portion of the production process which the machine presently cannot perform and is, in effect, the residual of the first relationship. This second relationship defines the employee's "job." The third one, Man-machine, describes primarily a maintenance function (role-, rather than job-relevant) and, by that token, will be excluded from the discussion at hand. To the degree that advanced technology, as we define it, exists in the elements of sophistication of material and complexity of machines, the machine-product relationship predominates in the production process—necessity of human intervention is minimal or nonexistent. To the degree that sophistication of technology exists in the control feedback process we can talk about continued, albeit diminished, inclusion of the Man-product relationship. Even though the employee needs to continue to act on the more sophisticated feedback, his "job" (the Man-product relationship) becomes simpler under more complex feedback systems. He needs not wonder or wait under these conditions; the evaluative information comes quickly, unambiguously, and correctly via a direct process in addition to his own senses.

Role enlargement. It should be clear then, that the more sophisticated the Machine-product dimension, the more limited the Man-product relationships. Since the Man-Product relationship is what we traditionally think of as the "job," what happens to the worker under conditions of sophisticated technology (i.e., minimal Man-product relationship)? One option, of course, is that his "role" (defined as a set of rules and expectations from the employee as well as the organization, which direct all of his occupational or "at work" behaviors) diminishes and disappears with his "job"—that is, he either is displaced or continues working at the "nonjob" until he retires or quits. Another option might be that the worker's role enlarges as his job diminishes—that is, the role enlarges vertically as the worker takes, or is given responsibility for

---

[1]Although man may also be the object changed, this does not concern us here.

production supervision, quality control, and maintenance supervision and comes into contact with more members of the organization to get things done. His role, then, becomes more complex and more demanding as his "job" becomes simpler. This involvement via "vertical job enlargement" (in the specific context used here, clearly "role enlargement"), is exactly the reverse of the kind of involvement included in "horizontal job enlargement" where the worker undertakes more, rather than less, of the Man-product functions—for example, from auto assembly line worker to coach builder.

Hypotheses[2]

The main hypothesis in this study is that work group acceptance of planned social system change in the direction of more participative and responsible activities (e.g., Likert's "System IV") will be enhanced in subunits with initially high or complex technology. This acceptance is increased as a function of the amount of situational constraint applied by the technology and the degree to which prior technological change can be considered a continuing source of behavioral constraint, and of "unfreezing" in the lifespace of the work group.

This hypothesis is derived from the evidence available and the theoretical notions generated in the previous chapter on studies and speculation to date. As a flow, the model is presented in Figure 1.

Figure 1

DIAGRAMMATIC FLOW OF THE GENERAL HYPOTHESIS

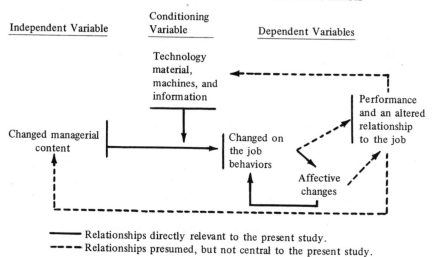

——— Relationships directly relevant to the present study.
- - - - Relationships presumed, but not central to the present study.

[2]Since the present study is essentially exploratory in nature, the hypotheses and predictions in this section are expressed in terms of working hypotheses rather than in specific null hypotheses and their alternatives.

The model implies three things immediately. First, it is postulated that new behaviors can effect changes in attitudes and that the new attitudes act to maintain the new behaviors. Second, it is postulated that the continued existence of the technology as a filter or conditioning variable acts to maintain the technology's constraint forces on behaviors. If the technology is sophisticated in the sense used here, those constraining forces will be in the direction of greater work group autonomy, worker discretion, and worker responsibility and will be considered facilitative constraints to the social change attempt. The presumed effectiveness of the continued existence of the technology implies that more direct management technologies for changing social systems (e.g., T-groups) may be less efficient, not because they are less constraining or compelling (exhortation, or pursuasion attempts are, however), but because they are not a continuous force for new behavior. Third, the primitive nature of the mathematical assumptions in this causal model must be recognized. The simple additivity of saying that within the technological classification, one unit of material complexity is equal to one unit of equipment complexity is equal to one unit of evaluation complexity is assumed because of the exploratory nature of the definition of the model and of the causal paradigm. There is no *a priori* reason for anything other than a straightforward combining of the components of technology or more closely specifying the causal influences. In fact, the analysis will involve assessing the conditioning effects of each of the three technological factors separately, as well as combining them.

Possible Confounding Elements in Measuring Technology

In classifying technologies as more or less complex for the purpose of providing a measure of a conditioning variable such as is presented here, two elements intrinsic to technology itself require consideration. The first, effectiveness of the technology, can and should be assessed. The second element, the debilitating, disruptive effects of change to more complex technology may not lend itself to control although it should be recognized.

Effectiveness of a new technology must be assumed in the model described above; the technology must do well that which it is supposed to do. If, for example, an "improved" technology is instituted which cannot physically provide more and/or better output, or the same output with fewer human interventions, it is not effective and the hypothesis above would not be expected to hold. It would naturally be assumed that the technology variable is truncated; that is, most complex technologies measured would be effective inasmuch as those which were ineffective would have been discontinued. It cannot, however, be concluded definitively that this is the case because some few technologies studied may still be within a testing or trial period, or may be recognized as ineffective but are retained for some other reason, such as prohibitive costs, in reverting to an older more effective system. In any event, such exceptions should be identified.

Disruption in technological change, the second element, may be associated with the effectiveness of the technology insofar as an intrinsically ineffective technology recently undertaken may be revealed only in disruption caused by employees' inability to adjust to that change within a reasonable period. The effect of disruption in technological change in the model of conditioning effects described above is considered basically facilitating if that disruption (or unfreezing) is of brief enough duration to allow for post-change stability. What this period is, or should be, depends primarily upon the extent and pervasiveness of the technology and the change. In general, we may assume that a start-up period (the period preceding improved production) should not be more than six months. Periods less than this will be considered facilitating (unfreezing); periods longer will be considered debilitating. Another source of debilitating side effects would be the effects on employee sense of security or competence. Generally it will be assumed that a change is disruptive in this sense if the members of groups experiencing technological change are older women (who are less mobile and who have relatively limited skills), or younger members of unions with strong seniority-rights contracts. In both of these cases, the employees involved would be expected to feel insecure with regard to the continuance of their jobs or their ability to adjust to new work.

Figure 2, below, presents the expected conditioning effects of complex technology on efficacy of social change attempts, controlling for the contingency effects of efficacy and disruption.

Figure 2

CONTINGENCY EFFECTS OF THE CONDITIONING
VARIABLE, TECHNOLOGY, IN PLANNED SOCIAL CHANGE

		Efficacy of Complex Technologies	
		High	Low
Disruptive side effects of technological change	Low	Best social change effects 1	Low social change effects 2
	High	3 Low social change effects	4 Worst social change effects

Cell 1 is expected to predominate in frequency of complex technology occurrence in the present study. Cells 2 and 3 are expected to have low frequency of occurrence; and cell 4 should have very few, if any, occurrences. By that token, cases selected for the main analysis will include only those in cell 1, unless enough cases are present in the other cells to warrant analysis.

Working Hypothesis

Since the general hypothesis has been stated above, it will not be repeated here. The main working hypothesis derived from the general hypothesis is as follows:

> Work groups with more complex (i.e., higher) technologies will change more within the measurement period and more permanently as well in the areas of perceived supervisory and peer (work group) leadership, work group process, and satisfaction,[3] than work groups with less complex, or lower technologies.

Specific Predictions

First of all, it has been suggested that technology in and of itself can have a direct effect upon work group autonomy and responsible, participative activity patterns. It, therefore, seems useful to examine the degree to which supervisory and work group leadership, and work group process is a function of technological sophistication prior to the onset of any planned social system change effort.

> *Prediction 1.*   Initial levels of supervisory and work group leadership and work group process are expected to be significantly related to the sophistication of initial technology. The level of these scores are expected not to be larger than scores derived for later periods of measurement subsequent to the onset of social change attempts.

It is hypothesized that following the onset of a social change effort in the direction of more participative management, most supervisors will attempt to become more supportive, more participative, more helpful managers. In groups with initially more sophisticated technologies, these behaviors on the part of the supervisor will better match some of the situational constraints of that technology than either the old style of supervision did in the same groups or the new style of management in groups with more traditional technologies. It is further hypothesized that in groups with higher technology, because of this better fit, work group leadership (i.e., more intra-group cooperation,

---

[3]These concepts, or abstract terms, will be operationally defined in Chapter III.

decision-making, autonomy, and responsibility) following the social change efforts on the part of others external to the groups, and the supervisor will be manifest to a greater degree than in groups with more traditional technology. These combined supervisory and work group leadership changes measured subsequent to the change effort will culminate in greater work group autonomy in the period of subsequent measurement.

Although it has been hypothesized above that following a planned social change attempt, supervisory leadership, as an initial variable in the chain of social change, will change more rapidly than work group leadership and group process, it is not expected that the measurement periods used in the present study will be short enough to capture this lag. It will be predicted instead that all three of these variables will change from the initial measure (prior to the change effort) to the second measure (subsequent to that change attempt) as a function of technology. Although difference in lag is not expected to be absolute, it may be evidenced in differences in strength of relationships between technology and supervisory leadership as causal variables measured at time one, and time two; work group leadership measured at times two and three.

Prediction 2a. Greater change, measured in terms of more pronounced shifts in supervisory and work group leadership over time, will be manifest for groups with more sophisticated initial technology.

Prediction 2b. Work group leadership, time three, should be found more causally dependent upon initial technology and work group leadership time two, than it will to prior supervisory leadership.

It is implied in Prediction 2b, that peer leadership and group autonomy between measures two and three will become more self-sustaining via the reinforcing effects of more favorable attitudes toward those behaviors. That is, technology may also not account for as much variance in peer leadership time three as in time two measurement because attitudes toward peer leadership have become more favorable and act as an additional reinforcement for time three behavior. The Festinger theory of attitude change (Festinger, 1957) provides a useful predictive model in this regard:

(a) Time one: the work group exercises little leadership.
(b) Time two: sophisticated technology creating a favorable situation for the exercise of group leadership advocated by the change agent provides for greater success in the social change effort—group members are behaving in the new ways because it better fits the situation, not because they "like it better" in some absolute sense.
(c) Time three: having experienced the better fit between the new ways

of behaving and the technological situation, group members are faced with a dilemma; their favorable and early socialized values and attitudes toward more traditional management styles are in conflict with the favorable experience of group leadership at time-two. The constraints of the situation have not and do not lend themselves to reversion to the behaviors consonant with the older values and attitudes. Thus, new attitudes, consonant with the new behaviors, are likely to be established between time-two and time-three measurement as a function of the reduction of cognitive dissonance. These new attitudes tend to reinforce subsequent and perseverative behavior of the new order at time-three.

If it can be established that consonant behaviors and attitudes are more likely to be affected by social change efforts in groups with more sophisticated technologies than in groups with lower technologies, then it seems reasonable to postulate that these changes will also be more permanent.

Prediction 3a. Satisfaction with the work group, time three, is expected to be causally dependent upon work group leadership and autonomous group process, time two.

Prediction 3b. The relationships between time two satisfaction with the work group, and time two peer leadership and group process are expected to be lower than their time one and time three counterparts where technology is sophisticated.

III

METHODOLOGY

Site and Sample

This research was carried out on a population of approximately 300 work groups (involving some 3,800 individual respondents) in two organizations. These two organizations were surveyed three times over a period ranging from 12 to 24 months. As part of a larger study conducted by the Institute for Social Research (ISR) at The University of Michigan, these respondents completed paper and pencil questionnaires dealing with job-related matters, including practices of the supervisor and work group, perceptions of organizational activities, individual motivation, and satisfactions with various aspects of the work environment. Questionnaire measurement was made prior to any planned social change activity, and then at least once subsequent to its outset. In the insurance company site, average response rate was 84 per cent, over all administrations, while for the refinery, average response rate was 76 per cent.

Insurance company. One organization was a medium-sized mutual fire, casualty, and life insurance company headquartered in the Midwest, with several regional offices throughout the United States. Survey involvement and planned change efforts in this company were gradual, with four regions plus the home office surveyed initially in Winter, 1966. Organizational development activities were undertaken in two of those regions and in the home office shortly thereafter. During the following Winter, 1967, these five previously measured units were resurveyed. Organizational development work, or planned social change efforts were undertaken in all sites following this last series of measures. Finally, these regions and home office were surveyed again, roughly a year after the prior ones described. Data from all nonsupervisory work groups in these offices were used in the present study, after the exclusion of some 35 sales groups (adjusted N groups = 150).[4] Although it would have been interesting to include sales groups in the sample, it was impossible to do so methodologically because all of them had the same technology. Such

---

[4]Nonsupervisory here means groups of people who have no formally assigned subordinates, hence the members of these groups can be composed of fairly high level technical or professional people, as well as lowest level production employees.

lack of variance in the conditioning variable would have made analysis of these data impossible.

Refinery. The other company was a large, modern oil refinery located in the South. This refinery has three major divisions: manufacturing, maintenance, and administration. The entire plant was surveyed during Spring, 1968. The manufacturing division was surveyed again 100 per cent during the Fall, 1968, while the remainder of the plant was sampled 15 per cent. Between the first and second administrations of the questionnaire the organizational development activities were begun. Finally, the entire refinery was surveyed again in Spring, 1969. The data used in the present study once again included all nonsupervisory groups (N groups = 156), after the exclusion of some additional 50 construction crews which would have received the same technological evaluation.

Table 1 presents the essential nature of each organization and the periods of measurement available.

The Independent Variable: Social System Change Program

Within each organization a relatively uniform organizational development program was undertaken. In all cases, this program involved an attempted change in management values and behaviors in the direction of Rensis Likert's "System IV" (Likert, 1967). "System IV" involves a theory of organization and management in which high value is placed upon total organizational commitment to joint decision-making, participation, openness, trust and confidence, mutual influence, and the sharedness of organizational goals and mission. The change program included an ongoing and general communication and exhortation of these values by management in both organizations. Additional change activities (e.g., skill training, or sensitivity training) were undertaken in each organization, but the details of these activities varied between them. Within each organization, however, this social system change program, the independent variable in our model, is constant throughout the analyses reported herein.

Specifically, change activity or organizational development was undertaken along the following lines:

(a) High-ranking company officials met for an orientation to "System IV" management by representatives of the Institute for Social Research.

(b) Company began introduction of the study to employees, immediately prior to the initial survey. This introduction usually contained some information regarding the avowed purpose of the study and the University's part in it.

(c) Initial survey, administered by ISR employees.

Table 1

THE TWO ORGANIZATIONS MEASURED:
SPECIFIC MEASUREMENT PERIODS AND SAMPLE PROPORTIONS SOUGHT

Organization	Measurement Periods and Sample Size									
	1966	1967				1968				1969
	Wint.	Spr.	Sum.	Fall.	Wint.	Spr.	Sum.	Fall	Wint.	Spr.
Fire and Casualty Insurance Company Four Regions plus Home Office (N groups = 150)	100%	–	–	–	100%[1]	–	–	–	100%	–
Petroleum Refinery (N groups = 156)	–	–	–	–	–	100%	–	15%[2]	–	100%

[1] Two of the four regions included in Winter, 1967 had no social change prior to this measurement as did the other two and home office, but were involved in social change subsequent to it.

[2] Some 54 groups in the refinery were surveyed 100% at this time. This amounted to the whole of one of the three divisions, plus several separate groups. The remainder of the plant was surveyed using a simple random sample of all remaining employees.

(d) Institute for Social Research change agents would begin to work with internal company change agents assigned by the company to familiarize these company people with "System IV" in greater detail.

(e) After data by work group were available, the ISR change agents and company change agents would make themselves available as consultants to individual supervisors who wished help in using the data they were provided. This help generally took the form of change agents' joining in discussions with a supervisor and his subordinates regarding their own data. This attendance by change agents was very systematic at the top of the organizations, and less so further down the hierarchy.

(f) The change agents, on the basis of the survey data and other observations, then proceeded to develop training programs for use with various levels and groups within the company. These training programs, both structured and unstructured, were usually carried out by the change agents involved in the planning.

(g) Throughout all of this the companies undertook high frequency, if not intensive, publicity or educational campaigns regarding the overall program and its goals and progress. These campaigns took the form of articles in company newspapers and magazines, memoranda to management, bulletin board circulars, and verbal utterances by management when in contact with employees at the work place. In the insurance company, the two regions which did not experience the initial change attempt following the first survey were exposed to the initial magazine articles which were company-wide, but did not receive the other in-company communications until after the second survey.

The cycle (e) – (f) – (g) was repeated following each survey.

Since there was no reliable way to distinguish quantitatively the intensity of planned change efforts among groups, the development programs will be considered constant across groups within each company. It is, however, anticipated that more variability in intensity is manifest between the companies then exists within them, so by that token, all analyses will be undertaken separately for each company.

Measurement of the Conditioning Variable: Technology

Measurement of the level of work group technology was accomplished using the model of technological classification described in Chapter II as the basis. That is, an instrument was built which attempted to measure the degree of material, hardware, and informational sophistication. The method used as the operationalization of these constructs was that of structured judgments of work groups or functions on several scales by a small number of administrative and management people within the organizations.

An original instrument of 16 items or scales was constructed in a form roughly similar to the more familiar job description rating procedures used for setting wage rates in industrial organizations. Each scale was arrayed horizontally across the top of an 8½ X 11 inch sheet. A brief stem was printed above the scale which included from two to seven rather lengthy alternative statements arrayed on an unnumbered continuum. Below the alternatives scale on each page were 11 rows of response spaces so that a judge could evaluate up to 11 groups or functions on each page. Where a judge was to rate more groups, additional pages of the same scales were provided. Each judge evaluated all the groups he was asked to rate for each scale before going on to the next scale or item. Items originally included consisted of five scales measuring sophistication of input, three scales measuring sophistication of machines, five scales measuring the sophistication of feedback information, two scales measuring the effectiveness of the technology if it were less than five years old, and one scale measuring the time necessary to de-bug the system if it were less than five years old.

This original instrument was pretested in a small company manufacturing glass products (N groups = 23), using three judges who evaluated all groups on all scales. The judges were rather highly placed technical people (one production planning engineer, two industrial engineers, one of whom was also the comptroller) who were familiar with all operations in the plant. Each judge was provided with a set of instructions which included examples of evaluations of familiar tasks or functions to serve as benchmarks in the rating of actual groups they were about to undertake. These judges, and all judges subsequently participating in evaluation of groups in the test sites, were given these instructions and were also instructed to rate the groups in terms of how the technology was *as of the time of the company's initial survey*. The task, on the average, took the pretest judges less than an hour.

The data obtained from this pretest were evaluated in several ways. First, all items for all judges were intercorrelated and items that clearly did not relate to their mates within the construct combinations, or related so highly as to suggest almost total redundancy, were removed, and the individual judges were visually compared for general agreement on each individual item. The 13 items remaining after this initial analysis (ten content items, three control items) were ultimately retained in the test instrument. This instrument is included in Appendix A. Following this analysis, the responses of all judges were averaged for each group for each item retained from the first analysis. The intercorrelations among the average item scores per group on the content items (i.e., those items actually to be included in the "Input," "Throughput," and "Output," technology indices) were submitted to a Smallest Space Analysis (SSA) (Guttman, 1968; Lingoes, 1965, 1966; Lingoes, Roskam and Guttman, 1969). The SSA revealed that the items intended to measure each of the three concepts did in fact cluster together in two dimensions which fit the data extremely well (Coefficient of Alienation for Euclidian Two-space = .023). It was found that the input and throughput clusters were relatively close together with the output cluster quite distant from them. The proximity of input and throughput clusters is not surprising since it may be hypothesized that the development of automatic machines depends on standardization of relevant raw materials. Figure 3 presents the two-space solution derived from the SSA. The variables in Figure 3 are numbered in accordance with the variable numbering in Appendix A.

Figure 3

SMALLEST SPACE ANALYSIS SOLUTION,
IN 2-SPACE, FOR TEN TECHNOLOGICAL CONTENT ITEMS[a]

Pretest Data (N Groups = 23)

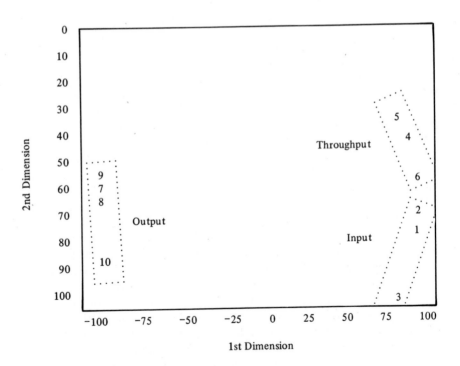

[a]For variable descriptions, see Appendix A.

This Smallest Space Analysis justifies combining item scores per judge into mean score indices for the input, throughput, and output concepts. SSA results are here considered as indicator of factorial validity, an indication that the technological instrument is a pure measure of the three concepts. Visual examination of the pretest groups evaluated and the average scores obtained suggested that those groups which, in the researcher's estimation, should have obtained a given ranking on a given technological index, did in large part do so. This is considered a rough indication of convergent validity.

These by-judge indices were tested for interjudge agreement using Kendall's Coefficient of Concordance, corrected for ties. Table 2 presents the results of this test. These results show reasonably good inter-rater reliability, or agreement among the judges—quite good, in fact, considering the experimental nature of the instrument and the rather alien character of the task for the judges.

These three, mean-score indices (plus two others, with small modifications described below) are the measures of the conditioning variable, Technological Sophistication, used in the two test sites described earlier. Not all the work groups were classified by technological sophistication; some had to be omitted. The omissions occurred for several reasons. First, direct production technology (referring here to the specific mission of ancillary staff groups, as well as to the primary production mission of lowest level line groups) is not ordinarily or easily applied to high- or middle-level management groups. Second, in some cases a great many groups (e.g., as much as 20 per cent in the insurance company) would have been given the same technology score since they performed the same function within the same technology. Hence, such groups where they appeared in large numbers were omitted. Third, groups or functions were omitted when the judges evaluated the groups as having ineffective technologies, or technologies so new as to preclude evaluation of their effectiveness. Since the number of groups so rated was very small (5 per cent of the total for the insurance company and 12 per cent of the total for the refinery), these groups were simply omitted rather than set aside for a separate analysis. This final omission of groups, in essence, controls for the postulated confounding effects of efficacy and of disruptive side effects in technology discussed in Chapter II. The final N of groups used in the test analyses, following this reduction were: N = 142 for the insurance company, and N = 140 for the refinery.

Calibration of the Technological Classification Form was undertaken for both test sites with varying degrees of success. In neither case was it possible to obtain multiple judgments of groups in any number. This lack was due to the large number of groups to be evaluated and the small number of potential judges available. The task, as it turned out, was tedious, demanding, and took quite a bit of a judge's time. Although the judges' cooperation was very high, it was an extremely large task just to get each selected group evaluated by one judge. Some multiple judge data per group were available, however, which were as follows. In the insurance company, two of eight participating judges were able to rate the same 11 functions (these are not work groups or shift groups on the same machine but are sets of duties carried out by groups of various sizes and compositions in each region of the company). The Spearman rank-order correlations (ρ) between the two judges over the 11 functions for the three indices are as follows: Input Index $\rho = .44$, $p < .10$; Throughput Index $\rho = .01$, N.S.; Output Index $\rho = .86$, $p < .01$. It is unfortunate that the interjudge agreement is so low for the Throughput Index, but it is observed that the insurance industry, unlike manufacturing, is not basically mechanical.[5] That "hardware" classes of technology were less suitable for white-collar organizations was noted in Chapter I.

Smallest Space Analyses similar to that made in the pretest site were undertaken with the insurance company data and the refinery data. The SSA in the insurance company produced a replication of the one obtained in the pretest site; input and throughput clusters were very close together, and the output cluster very distant. The results of the SSA in the refinery closely parallelled those obtained in the pretest with two exceptions. First, input and throughput clusters were not in close proximity, which suggests that, unlike the glass products factory used in the pretest where different types of glass are worked in different sets of machines, the pieces of equipment in a refinery (crackers, stills, tanks, etc.) are used for all the standardized yet different inputs, thus establishing more independence between input and throughput. Second, the absence of an "input" item from the appropriate cluster and its appearance in the output cluster was noted and examined. The item in question—knowledge availability for input (Item 3 in Appendix A)—clustered with the item dealing with the supervisor as a source of feedback (Item 7). This reveals that, unlike the pretest site where knowledge of input evidently resides in the operator, the knowledge regarding input in refinery operations resides instead in the supervisor. With this in mind, two additional technological indices were constructed: (a) Input Index without the knowledge item (Item 3), and (b) Output Index without the two supervisor items (Items 7 and 8 in Appendix A), which more closely match the close clustering of Items 9 and 10 in the unmodified Output Index.

Five measures of Technological Sophistication resulted from the original instrument, the pretest, and the calibration of the test sites with the pretest site. These five measures are as follows:

> (a) A three-item mean score index of input sophistication (Items 1, 2 and 3, Appendix A).

---

[5] As will be shown in Chapter IV, the Throughput Index did produce meaningful and interpretable results in spite of this lack of reliability evidenced for a portion of the insurance company data.

(b) A two-item mean score index of input sophistication (Items 1 and 2).

(c) A three-item mean score index of throughput sophistication (Items 4, 5 and 6).

(d) A four-item mean score index of output sophistication (Items 7, 8, 9 and 10).

(e) A two-item mean score index of output sophistication (Items 9 and 10).

These five indices of Technological Sophistication were subsequently constructed for both test sites. The resulting data from each site were examined separately in assessing the degree to which obtained distributions approximated normal distributions. This was done by computing the usual statistics (mean, standard deviation) and estimates of skewness and kurtosis as well for each index in each site. The results of this revealed that Index "b," the two-item Input Index, better approximated a normal distribution than did "a," the three-item Input Index. Also revealed was that Index "d," the four-item Output Index, better approximated a normal distribution than did Index "e." The Throughput Index (Index "c") had no alternative, but approximated a normal distribution reasonably well. Means, standard deviations, skewness and kurtosis estimates for Indices "b," "c," and "d" for both test sites are presented in Table 3. These three indices were those used in the test analysis described below.

Table 2

KENDALL CONCORDANCE COEFFICIENTS, W, AMONG
JUDGES OVER GROUPS FOR THE THREE TECHNOLOGY
CONTENT INDICES, CORRECTED FOR TIES.

(Pretest Data; N Groups = 23, N Judges = 3)

	$\dot{W}$
Input Index	.676**
Throughput Index	.756***
Output Index	.667**

**$p < .01$.
***$p < .001$.

Table 3

MOMENTS OF THE DISTRIBUTIONS OF THREE INDICES OF
TECHNOLOGICAL SOPHISTICATION FOR THE TWO TEST SITES

Sophistication of Technology Indices	Mean (E^1)	S.D. (E^2)	Skewness (E^3)	Kurtosis (E^4)
		Insurance Company		
Input (Index b)	3.45	0.99	−0.13	−1.19
Throughput (Index c)	1.42	0.85	−0.29	−0.94
Output (Index d)	2.66	0.38	0.37	−0.34
		Refinery		
Input (Index b)	3.38	0.94	−0.15	−0.68
Throughput (Index c)	2.41	0.67	1.00	0.19
Output (Index d)	2.48	0.39	0.17	−0.84

Measurement of Dependent Variables: Work-Related Behaviors and Attitudes

The survey instrument administered over time to the two companies included over 100 items, some of which were used as single item estimates of constructs or concepts; others were consolidated into mean score index variables as measures of concepts. From this large number of variables, ten were used in the analysis of the present study. These variables can be categorized as falling into four classes:

(a) Perceptions of supervisory leadership in the areas of support for subordinates, goal emphasis, work facilitation, and interaction facilitation.
(b) Perceptions of work group, or peer leadership in support, goal emphasis, work facilitation, and interaction facilitation.
(c) Perceptions of work group activities.
(d) Individual member attitudes regarding satisfaction with the work group.

The specific variables and questionnaire items are spelled out in more detail below. These ten measures, then, form the basis of measuring the dependent variable—the degree of acceptance to the planned social system change in each organization by work group. To the degree that group means on these items increase over time, it is postulated that the social system change attempt is successful and resistance is low. This postulate is conditional with regard to certain of the variables as will be described below. The majority of the questionnaire items below use the standard response alternative set which is a modification of the Likert Scale typical of those used in many organizational survey studies.

1) To a very little extent
2) To a little extent
3) To some extent
4) To a great extent
5) To a very great extent

Supervisory Leadership—Measured by the following four mean score indices:

Support—behavior which increases his subordinates' feeling of being worthwhile and important people. (Mean score index—three items)

In the surveys this was measured by the following questions:

How friendly and easy to approach is your supervisor?
When you talk with your supervisor, to what extent does he pay attention to what you are saying?
To what extent is your supervisor willing to listen to your problems?

Goal Emphasis—behavior which stimulates an enthusiasm among subordinates for getting the work done. (Mean score index—two items)

The surveys used these items to measure this aspect of his behavior:

How much does your supervisor encourage people to give their best effort?
To what extent does your supervisor maintain high standards of performance?

Work Facilitation—behavior which helps his subordinates actually get the work done by removing obstacles and roadblocks. (Mean score index—three items)

These items measured this form of behavior:

To what extent does your supervisor show you how to improve your performance?
To what extent does your supervisor provide the help you need so that you can schedule work ahead of time?
To what extent does your supervisor offer new ideas for solving job-related problems?

Interaction Facilitation—behavior which builds the subordinate group into a work team. (Mean score index—two items)

These items were used to measure behavior of this kind:

> To what extent does your supervisor encourage the persons who work for him to work as a team?
>
> To what extent does your supervisor encourage people who work for him to exchange opinions and ideas?

Peer (Work Group) Leadership. This was measured by survey questions and indices usually identical to those used to measure the manager's leadership. In this case, however, the questions are worded, "To what extent are (do) persons in your work group. . ."

> *Support* (Mean score index—three items)
> friendly and easy to approach
> pay attention to what you're saying when you talk with them
> willing to listen to your problems

> *Goal Emphasis* (Mean score index—two items)
> encourage each other to give their best effort
> maintain high standards of performance

> *Work Facilitation* (Mean score index—three items)
> help you find ways to do a better job
> provide the help you need so that you can plan, organize, and schedule work ahead of time
> offer each other new ideas for solving job-related problems

> *Interaction Facilitation* (Mean score index—two items)
> encourage each other to work as a team
> emphasize a *team* goal

Work Group Activities

> *Work Group Team Process* (Mean score index—three items)
> In your work group, to what extent is work time used efficiently because persons in the work group plan and coordinate their efforts?
>
> To what extent does your work group make good decisions and solve problems well?
>
> To what extent do you feel that you and the other persons in your work group belong to a team that works together?

Satisfaction

> *With Work Group*
> All in all, how satisfied are you with the persons in your work group?

Justification for assuming that the operational measures described above do in fact have reasonable reliability and validity comes from many sources.

In general, it was found that individual respondent distributions for the items used as measures of dependent variables described above, including those combined into indices, were reasonably unimodal and symmetric for both test sites. Thus, in a fundamental sense these variables at least roughly fulfill the metric measurements of normal frequency distribution necessary for subsequent analyses described below.

The supervisory and work group leadership indices were originally described by Bowers and Seashore (1966). In that paper, the eight areas of leadership were defined and their ability to predict organizational performance was demonstrated. The internal consistency of the index measures has been subsequently examined and the results of this analysis suggested only minor modifications (Taylor, 1971). These suggested modifications were effected in the present study.[6]

The work group activities measures are similar in content and form to items used in many previous studies (*cf.*, Likert, 1961). The three items included in the present study were intercorrelated for data obtained from the refinery sample and these were examined by Smallest Space Analysis. The results suggested association among the three items included in the "work-group team process" index, and independence among a remaining three items. It was, therefore, decided to combine the former three items, since they apparently added to a measure of a single concept.

Very early in the research work in organizations done at the Institute for Social Research, five dimensions of employee satisfaction were determined, by factor analysis, to provide adequate coverage of the concept of morale (Kahn and Morse, 1951). These five dimensions were ultimately found to be most efficiently measured with five single items determined to be the best estimators of each. At least, these five individual satisfaction measures have since that time been used as the basic measures of employee attitudes in organizational studies undertaken by the Institute. Among these five is satisfaction with the work group, which is the single item estimator used in the present study.

Measurement of Control Variables

Social Alienation. Following Blood and Hulin (1967), it seems useful to plan to control the obtained data for the effects of alienation from middle-class norms. The single most powerful measure of this alienation was found to be the primary residence and experience, urban or rural, of the respondents.

---

[6]Additional analyses reported elsewhere (c.f., Taylor and Bowers, 1971, Chapter VII) provide estimates of internal consistency reliability, and discriminant validity for these measures.

This finding is further reinforced by the results of Turner and Lawrence (1965). By that token, a measure of respondent background was included to test the effects of this factor on the dependent variables. Since it may be possible that women in their traditional cultural roles react to the enlarged white-collar jobs in a similar manner to the socially alienated in blue-collar work (Hoos, 1961), an exploratory control by sex of respondent was also undertaken. The item used to measure the background control variable is presented below.

While you were growing up—say until you were 18—what kind of community did you live in for the most part?

1) Rural area or farm
2) Town or small city
3) Suburban area near large city
4) Large city

Age and Company Tenure. These could also elicit differences in the receptivity of respondents to change inputs, either technological or social. Blau (1955) suggests, at least in civil service, that older or long-tenured workers (those not under initial job probation) are more likely to react to change flexibly and to be willing to deviate in the service of better production (pp. 199, 208). Bright, on the other hand, maintains that in the factory, automation produces negative attitudes and inflexibility in older, longer tenure workers (1958, p. 203). Since it was found that a high relationship between age and company tenure existed,[7] it was decided that only tenure would be used in a test specifically of possible effects on dependent variables. In light of this relationship, it was felt that tenure would adequately represent the effect of both variables. Because available data are not unambiguous on this point, no specific predictions were made regarding the direction of influence tenure may have, but this variable was tested by site for possible effects on the dependent variables. Items measuring age and tenure are presented below.

Into what age bracket do you fall?

1) 25 years or under
2) 26 years to 30 years
3) 31 years to 35 years
4) 36 years to 40 years
5) 41 years to 45 years
6) 46 years to 55 years
7) 56 years or over

When did you first come to work here?

1) Less than 1 year ago
2) Between 1 and 5 years ago
3) Between 5 and 10 years ago
4) Between 10 and 15 years ago
5) Between 15 and 25 years ago
6) More than 25 years ago

---

[7]In the insurance company sample, age and tenure were related r = .72 and in the refinery, r = .84.

Education. Converse (1964) has shown that education can have a marked effect upon the individual's ability to hold a set of beliefs (and concomitant behaviors) which form an interrelated and internally consistent value position. Since the planned social change attempt implies a consistent series of propositions, it could be hypothesized that less educated workers and supervisors could behave and believe one portion of the new management system without feeling any compelling demands to attempt to behave or believe the rest of the system presented. For example, it might be true that workers, and lower-level management, accept increased communication and greater frequency of meetings while denying that increased responsibility would also follow, or denying that increased responsibility would necessarily demand greater authority.

Mueller (1969, Chapter 8) in a study of labor force reaction to technological change, found that workers with higher formal education held jobs using more sophisticated technology. She also found that high formal education was associated with higher job satisfaction and greater acceptance of automation. She found no evidence, however, that sophistication of technology led to the higher educated worker's higher satisfaction; in fact, for a given level of technology, she found attitudes toward the machine equal for high and low education. Mueller speculates that higher educated workers are more satisfied and adaptable simply as a function of their education. She suggests that the less educated are less adaptable perhaps because they feel their lower skills will more likely lead to technological displacement. We can conclude that higher technology may actually require higher education, or that employers may simply set very high education requirements for hiring when modern technological systems are to be staffed because they want the "better outlook" and greater adaptability of the more highly educated. Mueller admits her data do not distinguish between these hypotheses.

Respondent education and its effect on the dependent variables as representing acceptance of the social change technique and its value system was tested. The measure of respondent education is presented below:

How much schooling have you had?

1) Some grade school   4) Completed high school
2) Completed grade school   5) Some college
3) Some high school   6) Completed college

Analysis Procedures

The analysis of results involved three phases: (a) a check of demographic variables for possible contamination of the dependent variables and (b) a test for interaction between the conditioning variables and interrelations among the dependent variables, and (c) a test to estimate causal priorities in the model.

The first phase amounted to testing the dependent variables, measured at the final administration, for possible confounding effects of the demographic variables described above. This control check involved testing for both univariate effects, using estimates of curvilinearity (eta); and multivariate effects, using multiple correlation and multiple regression techniques. Both steps were accomplished using the Multiple Classification Analysis (MCA) (Andrews, Morgan and Sonquist, 1967). The computer program used was the MCA 360-40 version at the Institute for Social Research.

The second phase of the analysis involved testing for possible interaction effects between the postulated conditioning variable, sophistication of technology, and interrelationships among the supervisory and peer leadership, and the group process variables. This was accomplished by intercorrelating the dependent variables while controlling for approximate median splits on the three major technological indices.

This test for interaction was originally planned as a prior step to analyzing the data by Path Analysis Techniques (Duncan, 1966). The interaction effects found were so strong and complex, however, as to rule out Path Analysis as a method of testing the model proposed in Chapter II. Since the hypotheses put forward in Chapter II are clearly causal in implication, an analysis method which would provide the most information and yet not be particularly error-prone was sought. Though correlations using change variables make sense intuitively in the present study, failure to note the dependence of such correlations on other underlying correlations may lead to errors of inference (Bereiter, 1967). For this reason, Path Analysis was considered as a less direct, but methodologically safer method, given that the conditions of linearity and additivity were met. Inasmuch as the latter conditions were not met, it was decided instead to test the causal model via zero order correlations among the variables, plus a simplified modification of the cross-lagged Panel Correlation Technique (Campbell, 1967, Pelz and Andrews, 1964; and Rozelle and Campbell, 1969). The cross-lagged technique was applied while controlling the data by the high-low splits in the technological variables which were found to affect relationships among the dependent variables. As a method of estimating causal priority, the cross-lagged method seems to avoid the problems encountered using change, or difference-scores, while at the same time being less complex to calculate and involving less stringent conditions than Path Analysis.

All analyses are based on group mean scores rather than on individual level scores for both theoretical and pragmatic reasons. Since tasks in the work group are usually related to one another, it is the technology of the total group that is of interest, and measures of the technology of individual jobs were simply not available or possible to obtain. With respect to the dependent variables, combining individual perceptions of supervisory and peer leadership, and work group process produces more stable measures of these constructs than would the individual perceptions alone.

All analyses are controlled for organization. Although in this case the planned social change attempt can be considered uniform throughout, it is recognized that the strategies for change within each organization varied one from the other. Therefore, each of the three phases of the analysis described above will be duplicated for each site. Furthermore, the insurance company sample will be divided into two parts for separate analyses as well. These two parts will be referred to as "test," and "control" sites. It was noted in Table 1 that the insurance company sample included four territorial regions plus the home office, all surveyed 100 per cent at three different times. It was also noted that two of the territorial regions plus the home office undertook planned change programs in participative management immediately following the first measurement, while the two remaining regions did not. Since the thrust of the present study is on technological effects in reaction to change programs, the two latter regions in the insurance company can act as "controls" to the two former regions and home office (the "test" sites) during the period between the first measurement and the second. Inasmuch as all four regions plus home office undertook the change program following the second measure, the control regions act as a replication between time two and time three to the test sites between time one and time two. Aside from the demographic checks on the dependent variables, which it was assumed would not be affected by the "control"-"test" conditions, all analyses were run separately for the refinery, the insurance company "test sites," and the insurance company "controls."

IV

RESULTS

This chapter presents the findings of the research together with minimal discussion of their meaning and implications. The first portion of this chapter deals with some preliminary analyses. The second reports an unexpected pattern of findings which was obtained and examines data relevant to a test of the global hypothesis. The third portion reports the results of the tests of the specific predictions to the working hypothesis.

Preliminary Analyses

Before testing the hypotheses and predictions of Chapter II, an examination must first be made of the possible confounding effects on the dependent variables Specifically, two questions are of interest: a) What effects do the demographic variables have on the dependent variables at one point in time? b) To what degree can the three technological variables be said to produce an interaction effect in the relationships among the dependent variables? These questions will be considered in turn.

Effects of Control Variables

A number of social characteristics or background variables which could possibly affect the absolute levels of dependent variable means were described in Chapter III. These effects were studied statistically by means of a computer program called Multiple Classification Analysis (MCA) (Andrews, Morgan, and Sonquist, 1967). The results derived from using MCA provide estimates of the effect of each predictor alone, the effect of each while controlling the effects of the others, and the effects of all predictors taken together.

The results of the MCA using the background items of a) tenure with the company, b) amount of education, c) respondent background (where the respondent lived until he was 18 years old—rural, town, or city), and d) respondent sex[8] as predictors are presented for the insurance company and

---

[8]Since the demographic variable respondent sex was reasonably distributed only in the insurance company, it was not included in the MCA for the refinery. It was known in

(continued on page 54)

refinery in Tables 4 and 5. The variables predicted to in this analysis include the ten basic questionnaire variables, (the dependent variables) measured at time three. The set of predicted variables included the three technological indices as well. These last three variables were included as a matter of interest merely to see if social characteristics were associated with them, and not for potential control. Tables 4 and 5 present the statistic eta-squared (η^2) as an estimate of variance accounted for in each dependent variable and technology variable, by each social characteristic alone, and beta-squared (β^2) as estimate of a predictor's ability to account for variance in a dependent variable while controlling for the other predictors or background items. Tables 4 and 5 also present the statistic R^2 as an estimate of variance accounted for in each dependent and technological variable, taking all background variables together.

Effects on the Dependent Measures. It is seen in both Tables 4 and 5 that in the case of the ten dependent variables, little effect is manifest by any or all social characteristics taken together. These four background items when taken together (average R^2) account for an average of only 5 per cent of the variance in the dependent variables for the insurance company (Table 4). The three social characteristic items used in the refinery account for an average of only 3 per cent of the variance (Table 5). Of all the social characteristics, education appears to have the strongest effects on the dependent variables. In the insurance company data, these relationships are all positive in direction, with the effects on supervisory leadership being the strongest. In the refinery, on the other hand, peer leadership variables are more strongly related to education than are the rest and where the relationships between education and supervisory leadership are positive in direction, those with peer leadership and group process are negative. Since these results of the MCA show that little variance is accounted for by those few items which revealed effect, and since correcting for these small effects could introduce new biases into the data (Blalock, 1964, pp. 86-87, 163), it was decided to leave the dependent variables unadjusted for tenure, education, respondent background, and sex. In light of the major interest here, not in specific indices, but in groups of indices—supervisory leadership, peer leadership, and group process—the lack of consistent effects of social characteristics within these combinations, or between the two organizations, is additional justification for not adjusting the dependent variables.

Effects on the Technological Measures. As for the effects of social characteristics on the technological variables, Tables 4 and 5 reveal a rather

---

(continued from page 53)
advance that the insurance company was composed of nearly 50 per cent women, while the refinery had practically none. In fact, the refinery was surveyed in such a way as to permanently code the data for all secretarial personnel into several separate groups which accounted for less than 3 per cent of the total number of groups used in the present study.

Table 4

ESTIMATES OF VARIANCE ACCOUNTED FOR (η^2), AND RELATIVE ABILITY TO
ACCOUNT FOR VARIANCE (β^2) BY SOCIAL CHARACTERISTICS, ON DEPENDENT
VARIABLES AND TECHNOLOGICAL VARIABLES

Insurance Company Data

Predicted Variables	Predictor Variables								R^{2a}
	Tenure		Education		Background		Sex		
	η^2	β^2	η^2	β^2	η^2	β^2	η^2	β^2	
Dependent Variables									
Supervisory Support	.02	.01	.05	.06	.02	.02	.00	.01	.05
Supervisory Goal Emphasis	.02	.01	.05	.07	.00	.00	.00	.01	.03
Supervisory Work Facilitation	.01	.00	.04	.05	.00	.00	.01	.00	.00
Supervisory Interaction Facilitation	.03	.01	.08	.15	.05	.02	.00	.04	.09
Peer Support	.05	.03	.05	.07	.00	.00	.00	.02	.07
Peer Goal Emphasis	.06	.04	.04	.09	.00	.01	.00	.05	.09
Peer Work Facilitation	.00	.00	.01	.08	.02	.04	.02	.08	.06
Peer Interaction Facilitation	.01	.00	.01	.09	.01	.02	.02	.09	.05
Satisfaction with Work Group	.04	.04	.00	.00	.00	.00	.01	.02	.02
Group Process	.01	.00	.01	.08	.01	.02	.01	.08	.05
Technological Variables									
Technological Input	.18	.06	.50	.28	.01	.00	.28	.03	.56
Technological Throughput	.05	.02	.12	.05	.03	.02	.07	.02	.14
Technological Output	.19	.11	.32	.07	.01	.00	.26	.11	.44

[a]The R statistic, and hence this R^2, is adjusted for degrees of freedom.

Table 5

ESTIMATES OF VARIANCE ACCOUNTED FOR (η^2), AND RELATIVE ABILITY
TO ACCOUNT FOR VARIANCE (β^2) BY SOCIAL CHARACTERISTICS,
ON DEPENDENT VARIABLES AND TECHNOLOGICAL VARIABLES

Refinery Data

Predicted Variables	Predictor Variables						R^{2a}
	Tenure		Education		Background		
	η^2	β^2	η^2	β^2	η^2	β^2	
Dependent Variables							
Supervisory Support	.00	.01	.05	.03	.04	.02	.02
Supervisory Goal Emphasis	.00	.00	.01	.01	.03	.04	.01
Supervisory Work Facilitation	.00	.00	.02	.03	.01	.02	.02
Supervisory Interaction Facilitation	.01	.02	.00	.00	.00	.01	.03
Peer Support	.02	.02	.00	.00	.00	.00	.03
Peer Goal Emphasis	.00	.00	.04	.04	.02	.01	.01
Peer Work Facilitation	.02	.01	.14	.12	.05	.00	.11
Peer Interaction Facilitation	.02	.01	.09	.08	.03	.00	.06
Satisfaction with Work Group	.04	.04	.01	.01	.01	.00	.01
Group Process	.04	.02	.10	.09	.03	.00	.07
Technological Variables							
Technological Input	.07	.02	.39	.34	.14	.01	.39
Technological Throughput	.01	.02	.05	.09	.01	.04	.03
Technological Output	.04	.03	.30	.35	.04	.01	.31

[a]The R statistic, and hence this R^2, is adjusted for degrees of freedom.

stronger pattern. The η^2 estimates for the input and output variables show a large amount of variance being accounted for by tenure, education, and sex in the insurance company, and by background and education in the refinery. Examination of the β^2 estimates for company tenure, background, and sex, however, reveals that relative ability to account for variance by tenure drops, and drops markedly, for respondent sex as well in the insurance company. Ability to account for variance by background drops in the refinery when the effects of the remaining two demographic variables are taken into account. The β^2 estimates for education, on the other hand, remain nearly as high as the η^2 estimates, suggesting that education does in fact have quite a large effect on the measures of sophistication of input in the insurance company, and on input and on output in the refinery. The actual distribution of responses showed different patterns of relationships for the two variables. First, education was negatively related to the sophistication of input for both organizations—that is, that respondent groups with high average education worked with input materials which were highly unstandardized. Second, education was found to be positively related to sophistication of output in the refinery—high education being associated with feedback modes where someone or something other than the immediate supervisor was the primary source of evaluative feedback. These results are not surprising since it seems quite reasonable to assume that an organization would select people with higher education to fill jobs where much discretion was required in *choosing* treatments for *unstandardized* inputs, and where they would work independent of supervisory evaluation. It might seem reasonable to consider adjusting these two conditioning variables (technology input, and output) for the effects of education, were it not for the fact that it was at least tacitly expected that education acts as a variable associated with technology and is not extraneous to it with regard to the dependent variables. In a national sample study of technological effects on the labor force, Eva Mueller has recently found that education is strongly related to technological sophistication of jobs held (1969, Chapter 8).

Social characteristics, taken together, are here considered a factor extraneous to the area of primary research interest in the case of the dependent variables (i.e., we are not interested in the effects of social characteristic as an independent predictor of dependent variables). Social characteristics should not, however, be considered extraneous in the case of the technology variables where the selection of one may be contingent upon the other. It is important to note that the input and output measures contain such a heavy component of education. This effect will not be adjusted for, but will be considered as the source of a possible rival hypothesis.

Interaction Effects

As described in Chapter III, the originally proposed analysis included a method of multivariate regression analysis which has strict requirements of linearity and additivity. In light of this, a test of the most crucial interaction effects—those between the level of technology variables and the relationship among the dependent variables over time—was undertaken. The importance of this test is as follows. If the effects of technology and the effects of prior levels of supervisory leadership, peer leadership, and group process act in a direct manner on subsequent levels of leadership and process following a planned change attempt, then we may say that technology and leadership act as additive components in determining the change process. If, on the other hand, technology acts on the *relationships* between levels of leadership and process prior and subsequent to change (for example, that the *level* of sophistication of technology acts to facilitate the *relationship* between peer leadership time one—before change efforts—and peer leadership time two—following change efforts), then we must conclude that technology acts not as an additive component with prior levels of leadership and process, but involves a different function altogether. Since this situation seemed likely, the test for interaction was undertaken.

Each of the three technological indices was divided on as close to a median basis as possible for each of the sites—refinery, insurance company test site, and insurance company control site. Pearson Product-Moment correlations among the dependent variables, time one-time two,[9] time two-time three were calculated for each of the median splits of each of the three technological variables, for each of the three sites—18 sets of intercorrelations altogether. The individual correlation coefficients within site, and technological index, transformed into Z's, were then matched in pairs for the high-low median split and compared for significant differences. This, then, provided a test for the question: Are the relationships among the dependent variables over time different for groups with high versus low sophistication on a particular measure of technology? Table 6 presents the summary statistics in answer to this question. The summary statistic in this case is the Z score obtained from a two-tailed Binomial test of the proportion of significant differences ($p < .05$) between matched correlation coefficients in each median split to the proportion of significant differences expected by chance alone at the 5 per cent level of confidence.

The findings in Table 6 clearly represent an interaction effect between most of the technological indices and the relationships among the dependent variables over time. The proportion of significant differences is uniformly greater for all sites and technology variables than would be expected by chance alone. The frequency (as well as strength) of these effects then rules out any analysis requiring additivity.

---

[9]The insurance company control sites were not tested for t_1-t_2 correlations since the test of interaction is for technological effect on planned change.

Table 6

TEST OF INTERACTION EFFECTS

Z RESULTS OF TWO-TAIL BINOMIAL TESTS OF PROPORTION OF SIGNIFICANT DIFFERENCES BETWEEN CORRELATIONS AMONG DEPENDENT VARIABLES FOR HIGH AND LOW LEVELS OF THREE TECHNOLOGICAL INDICES

	Technological Indices		
Organizations	Input	Throughput	Output
Refinery	4.11**	1.59	5.19**
Insurance Co. Test Sites	0.06	3.28**	2.17*
Insurance Co. Control Sites	11.24**	2.98**	0.23

$^{*}p < .05$
$^{**}p < .01$

Were it not for an unexpected patterning of results (to be described below) which led to a rather immediate and direct test of the major hypothesis, the interaction effect described above might have been subjected to the rather complex assessment of the function underlying the results with an eye to transformation of the data toward an additive model.

A Test of the Global Hypothesis

A Patterning of Relationships

Consistent differences within the interaction analysis. In reviewing the data obtained in the interaction analysis, it was noted immediately that the differences between the high and low groups for each site-technology matrix were not random. In fact, a pattern emerged even though the overall pattern might include few individual pairs of correlations with differences large enough to be significant. Since this is a rather complex finding, an example of what is meant would be useful.

An Example. In any given comparison of the high-low split for one of the three technology variables using either site, there is a similar matrix of inter-correlations. For example, supervisory leadership, time two (four indices) against supervisory leadership time three (four indices) is a part of each matrix; and it consists of 16 pairs (high-low technology) of intercorrelations for the 4 × 4 supervisory indices. In a typical case, it was found that the correlation coefficients for the high technology group on any technological variable were consistently different from the coefficients in the low technology group for the same variable and site. There were usually a few pairs of correlations where significant ($p < .05$) Z differences were manifest in the total (in the present example 16 pairs), and the remainder were consistently in the same direction although the individual differences were not significant. This patterning, however, would usually be consistent enough to produce a statistically significant two-tail binomial test for the proportion of differences in one direction being different from a 50 per cent chance occurrence. That is, in our example, comparing supervisory leadership time two, with supervisory leadership time three, 13 or 14 differences in the same direction out of 16—a frequent finding—was enough different from the expected chance occurrence of eight differences in the same direction, to produce a binomial Z significant at the .05 level of confidence.

"Connectedness," a possible explanatory variable. This sort of result suggested immediately that each of the three technological variables was having an observable effect on the patterning and strength of relationships among the dependent variables. It seemed that high technology could be affecting a facet of organizational behavior we might call (for lack of a better term) "connectedness" or system strength in organizational behavior—a concept derived from the existence of a network of relationships over time which suggests a certain interdependence or continuation of patterns of behavior, versus a lack of interdependence or connectedness.

In any event, no established method seems to exist at present to assess or measure the relatedness concept which presents itself in the present case in the form of differences in levels of correlations among the dependent variables over time. It was felt, therefore, that one way of looking at these findings would be to average the intercorrelations (using Z-score transformations) among the dependent variables for each of the time periods (t_1-t_1, t_1-t_2, t_2-t_2, t_2-t_3, t_3-t_3)[10] for each technology variable-site combination, and to use these averages as summary statements—coefficients of connectedness. These coefficients of connectedness could be compared by time period for the

---

[10]Not included in the simultaneous average relationship (i.e., t_1-t_1, t_2-t_2, t_3-t_3) were the relationships *within* the supervisory and peer leadership variables. It was known that such inclusions would only inflate all simultaneous averages about an equal amount.

high and low technology groups divided as above. Such a model provides a method of distinguishing two components—a within time connectedness $(t_1-t_1, t_2-t_2, t_3-t_3)$, and an across time connectedness (t_1-t_2, t_2-t_3). We might expect, that because of possible methodological problems, across time connectedness might be a better measure of system strength than within time connectedness since this latter component might suffer from position bias or a general halo effect to a greater degree than the former. It seems clear that we could expect greater association among questionnaire variables responded to by the same people at the same time than the same items responded to at different times.

The results of this computation of the coefficients of connectedness and the high-low comparison within each site and technology variable are presented in Tables 7, 8, and 9. Table 7 presents the coefficients of connectedness for the insurance company test site data comparing high and low median splits for each of the three technological variables. Tables 8 and 9 present the same data for the insurance company control site and refinery respectively. Although there is no way of assessing significance of average r's, the tables show levels of significance of the relationships if they were not averages.

Results for the insurance company: three technological indices. The results in Table 7 (insurance company test sites) reveal a pattern in the across time connectedness for the high technology half which is different from that of the low technology half, while little difference is evident between the high and low groups in the within-time connectedness. The high technology data for the input and throughput variables show little connectedness t_1-t_2, while some connectedness is shown for the output measure t_1-t_2. Between time two and time three (t_2-t_3) the pattern shows some reversal with input, throughput, and output all showing considerable interconnectedness. Low technology groups in Table 7 tend to show a pattern in reverse of that manifest in the high group. For the low group input and throughput measures, connectedness is nearly as high t_1-t_2 as it is t_2-t_3; and for output, connectedness t_1-t_2 is lower than it is t_2-t_3. The differences in the overall pattern of results with regard to the output measure in both the high and low groups is puzzling until additional data are reviewed. It will be remembered that the smallest space analysis of technological variables described in Chapter III revealed that the output cluster was quite distant from the other two technological clusters. This in itself might suggest that the output measure would behave differently from the other two. But it was also found in the insurance company that the output index was inversely related to the other two indices at a high level—the correlation between input and throughput was found to be $r = +.52$; between input and output $r = -.66$; and between throughput and output, $r = -.58$. Since in the insurance company it was also found that output was strongly and positively related to education while input was negatively related to education, there is little reason to be surprised by the reversed pattern in connectedness

Table 7

COEFFICIENTS OF CONNECTEDNESS COMPARING HIGH AND LOW GROUPS
IN THREE TECHNOLOGICAL VARIABLES

Insurance Company Test Site Data

Technological Variables	Time Period Comparisons				
	t_1-t_1	t_1-t_2	t_2-t_2	t_2-t_3	t_3-t_3
High Technology					
Input n=52	.61**	.14	.56**	.41**	.64**
Throughput n=37	.55**	.11	.45**	.31*	.38**
Output n=67	.42**	.25*	.46**	.31**	.49**
Low Technology					
Input n=39	.29*	.32*	.42*	.31*	.48**
Throughput n=54	.42**	.30*	.62**	.42**	.55**
Output n=24	.53**	.02	.65**	.43*	.44*

*Individual correlations this size would be significant at the .05 level of confidence.
**Individual correlations this size would be significant at the .01 level of confidence.

between output and the other two technological indices. As Converse (1964) has shown, higher levels of education are associated with greater behavioral and attitudinal consistency which may in the present case be evidenced in higher connected levels for high output and low input.

Table 8 presents the coefficients of connectedness for the insurance company control sites. While the insurance company test sites received the planned change program following time one, the control sites underwent the program following time two. If the connectedness coefficients reflect some

Table 8

COEFFICIENTS OF CONNECTEDNESS COMPARING HIGH AND LOW GROUPS
IN THREE TECHNOLOGICAL VARIABLES

Insurance Company Control Site Data

Technological Variables	Time Period Comparisons				
	t_1-t_1	t_1-t_2	t_2-t_2	t_2-t_3	t_3-t_3
High Technology					
Input n=36	.42**	.24	.27*	.14	.54**
Throughput n=30	.41*	.21	.25	.19	.37*
Output n=25	.69**	.24	.49**	.42**	.61**
Low Technology					
Input n=23	.70**	.35	.48*	.51**	.63**
Throughput n=29	.62**	.28	.49**	.36*	.71**
Output n=34	.44**	.26	.21	.17	.68**

*Individual correlations this size would be significant at the .05 level of confidence.
**Individual correlations this size would be significant at the .01 level of confidence.

systemic reaction to the planned change efforts, we should expect that the pattern of strength of connectedness found between time one and time two (t_1-t_2) in the test sites should be replicated in the control site data between time two and time three (t_2-t_3). This is precisely what obtains. Between time one and time two for both high and low groups in the control site there is some connectedness, but between time two and time three (t_2-t_3) the low technology groups show increasing connectedness for the input and through-

Table 9

COEFFICIENTS OF CONNECTEDNESS COMPARING HIGH AND LOW GROUPS
IN THREE TECHNOLOGICAL VARIABLES

Refinery Data

Technological Variables	Time Period Comparisons				
	t_1-t_1	t_1-t_2	t_2-t_2	t_2-t_3	t_3-t_3
High Technology					
Input n=78	.60**	.35**	.51**	.38**	.65**
Throughput n=64	.43**	.34**	.42**	.36**	.54**
Output n=84	.63**	.21*	.41**	.24*	.67**
Low Technology					
Input n=62	.55**	.23*	.31**	.24*	.50**
Throughput n=76	.67**	.25*	.41**	.26*	.63**
Output n=56	.49**	.40**	.40**	.32**	.45**

*Individual correlations this size would be significant at the .05 level of confidence.
**Individual correlations this size would be significant at the .01 level of confidence.

put indices, while connectedness increases for the output measure in the high technology group. These results strongly suggest that whatever the patterns mean with regard to strength and speed of behavioral change, technology (and perhaps education, following Converse) is affecting the reaction to change via strength of relationships between measurement periods. The within-time connectedness, time two-time two, it should be noted, exhibits greater similarity to the pattern of across time connectedness time two-time three in this site, than it does to the time one-time one within time connectedness in the insurance company test site.

Results for the refinery: three technological indices. Table 9 presents the connectedness data for the refinery site. As in the insurance company test site, differences between high and low technology are in evidence in the across time connectedness, but not in the within time connectedness. In the refinery, however, the overall pattern for across time connectedness differs from that of the insurance company. It is seen in Table 9 that connectedness is consistently higher for the high technology group input and throughput both between time one and time two, and between time two and time three. Connectedness is consistently lower in the low technology group for both indices and time lag periods. For the output measure, on the other hand, connectedness is higher for the low group for both t_1-t_2, and t_2-t_3. As was found to be the case for the insurance company, the output index was very different from the input and throughput measures in the refinery. The relationships between the three measures in the refinery are as follows: input-throughput, $r = -.20$; input-output, $r = -.30$, throughput-output, $r = -.60$. If, as was pointed out earlier, education is related to connectedness, the effects in the refinery data tend to be somewhat less supportive than those in the insurance company. Although it is seen in Table 9 that high output sophistication, and low input sophistication (both associated with higher education) have reasonable levels of connectedness, high sophistication of input and throughput have higher connectedness than their low counterparts. This suggests that technology is related to connectedness in addition to whatever effects accrue by both technology and connectedness being related to education.

Results Using a Combined, General Technology Measure

A commonality shown in Tables 7, 8, and 9 is the fact that within each company, the pattern of across time connectedness is the same for the input and throughput indices, and different for the Output Index. In an attempt to refine the analysis further, it was decided to combine the data such that groups with high input, throughput, and output scores could be compared with groups with low input, throughput, and output scores. This combination high-low design would reduce the number of groups in each category, but would still retain reasonable n's for statistical purposes.

Combination high-low technology scores using an optimization approach. Combination of technological indices was made in an attempt to optimize the scores of the three indices, either high or low. Because of the high negative relationships between the input and throughput measures and the output index, the probability of finding a reasonable number of cases with very high scores or very low scores on all three measures (an example of maximization) was very small. Empirically, there were no cases where this was possible in either site. The combination was, therefore, made in an attempt to obtain groups with moderately high or moderately low scores on all three technological indices. This was accomplished by separating first the high and low groups on the input index in an approximate median split. This was made the first step because input sophistication is conceived of as the primary element in sophisticated technology in general—most sophisticated machinery must be preceded by standardized, well-known raw materials. Second, within the high-low splits on input, groups were separated by a rough median split on the throughput index. Groups retained at this point in the sample were those which fell either high input-high throughput, or low input-low throughput. The throughput separation, rather than separation on output, was made following that for input, because it was felt (in terms of the reasoning in Chapter II) that for the present purposes, sophisticated channels of evaluative feedback, considered technological in origin, (the Output Index) would be more likely (i.e., to follow from) in systems with sophisticated input and throughput. Finally, then, the resultant high-high group was examined on the Output Index and those cases of lowest sophistication of output were discarded to the degree that a reasonable number of groups could be maintained (about 30 groups was considered an ideal minimum). The same procedure was used with the low-low groups with an attempt to remove the cases of highest sophistication of output. Although not totally satisfactory in an absolute sense, this procedure of optimization of technology scores seemed to meet the need of a combined score using all three indices.

A comparison of optimization to maximization. Since the absolute levels of the input and throughput scores for the groups in the samples described above were somewhat different from what might be expected by simply maximizing scores on input and throughput alone, such a maximization procedure was undertaken for purposes of comparison. In the maximization procedure, input and throughput measures were examined simultaneously for each group, and those groups with highest and lowest scores on both were drawn into the high-high and low-low sets without regard for output scores. This procedure had the advantage of clearly separating the groups on both measures. In the optimization procedure, on the other hand, some groups in high-high, and low-low categories had similar throughput scores since both groups were selected from the mode of the throughput distribution, rather than scores on either side because the initial separation on input had limited the degrees of freedom for selecting throughput cases, and sufficient throughput cases needed to be drawn in order to separate once again on the output measure.

Table 10 presents a comparison of the means obtained from the optimization and maximization procedures for combining the technology scores. Results in Table 10 reveal that, even with optimization, output scores for the low-technology groups in both sites are higher than the output scores for the

Table 10

COMPARISON OF MEAN SCORES ON TECHNOLOGICAL INDICES OBTAINED
FROM OPTIMIZATION AND MAXIMIZATION PROCEDURES

Method	Technological Variable		
	Input	Throughput	Output
Insurance Company Low-Technology			
Optimization	2.86	0.74	2.82
	*	*	**
Maximization	2.50	0.36	3.18
Insurance Company High-Technology			
Optimization	4.22	2.15	2.49
Maximization	4.21	2.19	2.40
Refinery Low-Technology			
Optimization	2.33	1.94	3.04
Maximization	a	a	a
Refinery High-Technology			
Optimization	4.23	2.02	2.50
	**	**	**
Maximization	3.89	2.95	2.15

[a] results using maximization were the same as those for optimization.
*t differences $p < .05$.
**t differences $p < .01$.

high-technology groups. It is also noted, however, that the maximization procedure obtains higher output scores for the low group and lower output scores
for the high groups than does the optimization procedure. For both procedures, input and throughput measures are lower for the low group, and
higher for the high group which, of course, is the primary purpose of the
combination using either procedure, given the nature of the output measure.
It would appear, then, that the optimization procedure does reasonably well
what it was expected to do. It will be used as the method of summarizing the

connectedness patterns change effects mean scores for the dependent variables, and the cross lag causality tests of the specific hypotheses.

Connectedness results using combined technology scores. Tables 11 and 12 present a comparative analysis of connectedness for high and low technology groups in the insurance company test site and the refinery, respectively. Since the number of groups would be reduced markedly for the insurance company control site and since it can be assumed that the control results will mirror those for input and throughput in Table 8, it was decided to omit the control site from the present analysis.

The pattern of connectedness coefficients for the insurance company in Table 11 replicates that found in Table 7. There is little across time connectedness between time one and time two for the high technology group, and much stronger connectedness for this group between time two and time three. The insurance company low technology group, on the other hand, had high across time connectedness during the first time lag period, and moderate connectedness for time two–time three using the combined technology measure. There was no difference between the two groups in within time connectedness.

Table 12, the high technology-low technology analysis for the refinery, also shows a pattern similar to that found for the input and throughput variables in Table 9—greater connectedness for the high group than low group in both time lag periods. In the present case, however, these differences are larger than those found using the individual technology indices. Some differences between the two groups were also revealed in within time connectedness time two—which had not been as strong in Table 9.

Level of education and combined technology measure. Since, as we have seen, education appears to be related to both the connectedness coefficients and the individual technological variables, it would seem necessary to check the combined technological measure used here (optimization method) for the possible effects of education in order to determine whether the connectedness scores in Tables 11 and 12 are a function of education instead of (or as well as) technology. The mean education score for the low technology group in the insurance company was found to be significantly greater than that for the high technology group (high technology education mean = 4.28; low technology mean = 4.88; $t = 3.80$, $p < .01$). This result might help to explain the high connectedness score t_1-t_2 and the moderate connectedness score t_2-t_3 for the low technology group in Table 11, but does not help to explain the high connectedness t_2-t_3 for the high technology group. This puzzling result will be reviewed again in the description of the cross lag analysis. The education scores for the high and low technology groups in the refinery were not significantly different from one another (high technology education mean = 3.97; low technology mean = 4.18; $t = 1.31$, n.s.). This result provides evidence that technology itself is affecting connectedness in the refinery to the practical exclusion of education.

Table 11

COEFFICIENTS OF CONNECTEDNESS
COMPARING HIGH AND LOW TECHNOLOGY

Insurance Company Test Site Data

Groups	Time Period Comparisons				
	t_1-t_1	t_1-t_2	t_2-t_2	t_2-t_3	t_3-t_3
High-technology n=29	.60**	.12	.53**	.35*	.44**
Low-technology n=22	.57**	.42*	.59**	.28	.57**

*Individual correlations this size would be significant at the .05 level of confidence.
**Individual correlations this size would be significant at the .01 level of confidence.

Table 12

COEFFICIENTS OF CONNECTEDNESS
COMPARING HIGH AND LOW TECHNOLOGY

Refinery Data

Groups	Time Period Comparisons				
	t_1-t_1	t_1-t_2	t_2-t_2	t_2-t_3	t_3-t_3
High-technology n=42	.72**	.27*	.53**	.33*	.71**
Low-technology n=26	.65**	.13	.32*	.14	.61**

*Individual correlations this size would be significant at the .05 level of confidence.
**Individual correlations this size would be significant at the .01 level of confidence.

Mean Change Over Time

The data so far have strongly suggested that differences between high and low technology are associated with differences in strength of relationships (or connectedness) among dependent variables over time and that these patterns of across time connectedness are related to the planned change effort (at least in the insurance company where the control data tend to establish this). The remainder of this portion of Chapter IV will explore the data most directly relevant to the prediction made in the general hypothesis—that high technology facilitates, not merely affects, the planned change process. The general high-low technological split obtained by the optimization procedure was used as a framework from which to test differences between mean scores on the dependent variables since it was felt that it provides a relatively stable and simple framework to test results, and one which allows testing the general hypothesis.

The mean scores for supervisory leadership, peer leadership, and group process for the high technology and low technology groups for time one and time three are presented in Table 13 for the insurance company data and in Table 14 for the refinery data.

The results in Table 13 are basically encouraging, although at first glance they would seem the reverse of what was expected. It was expected, for example, that the high technology groups, having greater situational constraint toward more autonomous, more responsible group structure might be found to have initially higher means in peer leadership and group process. This was not found to be the case in the insurance company. Nearly all (seven out of nine) mean scores time one (and time two group process, which was the first time this measure was used in the insurance company questionnaire) were higher for the low technology group than for the high technology group. Of these seven higher scores, six of them were significantly higher at the 10 per cent level of confidence or better. Thus, respondents in the high technology group initially saw the behaviors measured by the dependent variables less in evidence than did the low technology group. That this result is a function of education is not unlikely since, as we have seen from Table 4, education is at least somewhat positively related to the dependent variables. It will be recalled that the level of education was significantly greater in the low technology group than in its high counterpart.

The second expectation was that the means of the high technology group would increase more than those of the low technology group. It is clear in Table 13 that only one variable increased slightly between time one and time three, but it is also clear that nothing increased for the low technology group either. It was known before the present study was undertaken that the insurance company data in general had not shown an increase in participative management over time; it had in fact declined markedly. This decay has been

Table 13

MEAN SCORE DIFFERENCES ON DEPENDENT VARIABLES BETWEEN HIGH-TECHNOLOGY AND LOW-TECHNOLOGY GROUPS; AND WITHIN GROUPS OVER TIME

Insurance Company Test Site Data

Dependent Variables	High-technology			Low-technology			Differences	
	means		diff.	means		diff.	High$_1$-Low$_1$	High$_3$-Low$_3$
	Time$_3$	Time$_1$	t_3-t_1	Time$_3$	Time$_1$	t_3-t_1		
Supervisory Support	3.94	4.42	-.48**	4.04	4.68	-.64**	-.26*	-.10
Supervisory Goal Emphasis	3.82	4.01	-.19#	3.94	4.24	-.30	-.23#	-.12
Supervisory Work Facilitation	3.19	3.55	-.36*	3.34	3.50	-.16	.05	-.15
Supervisory Interaction Facilitation	3.65	3.69	-.04	3.70	4.03	-.33*	-.34#	-.05
Peer Support	3.83	4.15	-.32**	3.94	4.44	-.50**	-.29*	-.11
Peer Goal Emphasis	3.47	3.57	-.10	3.55	3.99	-.44**	-.42**	-.08
Peer Work Facilitation	3.29	3.26	.03	3.31	3.37	-.06	-.11	-.02
Peer Interaction Facilitation	3.30	3.33	-.03	3.31	3.72	-.41*	-.39#	-.01
Group Process	3.47	3.61[a]	-.14	3.42	3.59[a]	-.17	.02	.05

#p < .10.
*p < .05.
**p < .01.
[a]Group process was not measured t_1, t_2 data used instead.

Table 14

MEAN SCORE DIFFERENCES ON DEPENDENT VARIABLES BETWEEN HIGH-TECHNOLOGY AND LOW-TECHNOLOGY GROUPS; AND WITHIN GROUPS OVER TIME

Refinery Data

Dependent Variables	High-technology			Low-technology			Differences	
	means		diff.	means		diff.		
	$Time_3$	$Time_1$	t_3-t_1	$Time_3$	$Time_1$	t_3-t_1	$High_1-Low_1$	$High_3-Low_3$
Supervisory Support	4.16	3.92	.24*	4.10	4.14	-.04	-.22	.06
Supervisory Goal Emphasis	4.08	3.94	.14	3.85	3.75	.10	.19	.23*
Supervisory Work Facilitation	3.66	3.45	.21*	3.39	3.20	.19	.25	.27*
Supervisory Interaction Facilitation	3.92	3.64	.28*	3.43	3.23	.20	.41*	.49**
Peer Support	4.01	3.83	.18*	3.94	4.03	-.07	-.20	.07
Peer Goal Emphasis	3.74	3.55	.19*	3.48	3.44	.04	.11	.26**
Peer Work Facilitation	3.75	3.47	.28**	3.19	3.11	.08	.36*	.56**
Peer Interaction Facilitation	3.65	3.37	.28**	2.87	2.68	.19	.69**	.78**
Group Process	3.84	3.68	.16*	3.33	3.17	.16	.51**	.51**

*p < .05.

**p < .01.

variously attributed to forces external to the organization, to top management's reaction to these external forces, and to a change program where the real indicators must necessarily diffuse more slowly than the expectations and ideals which precede them (i.e., a case of the effects of increased expectations on perceptions of the state of the environment). In light of this general decline in strength of response in the insurance company, the test of technological effects on planned change must be an examination of relative rates of decay between the high and low technology groups. It is clear from Table 13 that the rate of decline is faster for the low group (the negative differences t_1-t_3 are greater). It is also apparent in Table 13 that whereas the low technology group was markedly higher in mean level of dependent variables at time one, by time three the differences between them are considerably less. At time three, the mean scores for the low technology group are still higher, but there are no significant differences between high and low technology groups. Where there were six out of nine significant differences between the high and low technology groups at time one, there are none at time three. This converging over time of originally different groups, where the high technology groups are decaying less rapidly than the low technology groups, provides evidence in favor of the original hypothesis (and, to a lesser extent, of the first prediction as well) that situational constraints of technology operate in the direction of more participative management. These results may also lend support to Mueller's finding that higher levels of education (the low technology group in this case) are associated with greater adaptability and more favorable job outlook. It seems not unlikely that situational constraints against change in the high technology group, and greater adaptability in the low technology group operated to provide the different rates of decline noted in Table 13.

Table 14, the mean differences for the refinery data, also supports the first prediction, and in this case more directly. At time one, the high technology group has significantly higher mean scores in two peer leadership measures and group process than the low technology group. In two cases out of nine at time one, the low group is higher than the high group, but, in the main, the latter is originally higher than the former. While eight of the nine variables show significant increase between time one and time three for the high group, not one of the nine shows increase which reaches significance levels for the low group. The high technology group improved at a faster rate than did the low technology group although they both improved over time. Finally, at time three, the high technology group had higher mean scores than the low technology group in all nine variables, seven of which were significant. These latter data are seen to confirm prediction 2a as well.

With reference to the data presented in Table 14, regarding prediction 2a, it should be noted that a potential confounding influence might be present. Table 14 shows that groups with high technology had higher initial leadership scores than groups with low technology. The data also show that in the groups where leadership scores are higher initially (i.e., high technology groups), the scores on these variables increase more over time than in groups where they were initially lower (i.e., low technology groups). A reasonable alternative explanation of these findings might be that the mere initial level of leadership scores influences subsequent shifts—that is, groups initially higher in leadership increase faster in leadership than groups initially lower in leadership. This effect was examined directly in the refinery data by assessing level of time three leadership results, while controlling for initial level by taking median splits for time one data over all nonsupervisory groups on each of the eight leadership variables. The results of this test are presented in Table 15. These data reveal not a "gain effect" on the part of groups with high initial levels of leadership, but, rather, a regression toward the mean for groups with initially high and initially low leadership scores. It would seem from this analysis that technology, by itself, accounts for the greater gains in leadership variables over time in the high technology refinery condition.

Results Regarding Specific Hypotheses

Some problems in the tests. The specific predictions 2b to 3b were written originially with the intent of using the technology measurement as a continuous predictor variable to the dependent variables. It should be clear from material presented earlier in the present chapter that technology could not be applied in this way using multiple regression techniques because of interaction effects found to exist between technology and the relationships between subsequent questionnaire variables (dependent variables) and these same variables measured earlier (considered predictors, along with technology). As noted in Chapter III, it had been felt that path analysis (the multiple regression technique) would have provided a reasonable substitute for a test of technology's effects on change or gain measures, and would also have provided a method of identification of causal paths over time. This would have been done by obtaining individual estimates of direct and indirect influence of technology and prior levels of the measures used to measure change on those subsequent measures while controlling for other causal variables in the model. Given that path analysis was proven inappropriate for these data, a somewhat cruder analysis plan was adopted. This included the high-low split on technology and the test for mean differences and change over time described in the last section; plus, as this section will show, an evaluation of the specific predictions, using cross-lagged correlations in establishing causal priorities.

The predictions in Chapter II took the form of implicit causal chains: the relative effects of technology, and early measures of the social variables, on subsequent measures. Technology was to be compared with prior measures of the social variables in assessing relative strength of influence on subsequent

Table 15

MEAN SCORE DIFFERENCES, OVER-TIME ON SUPERVISORY AND PEER LEADERSHIP VARIABLES. CONTROLLING FOR INITIAL LEVEL OF THOSE VARIABLES USING MEDIAN SPLITS

Refinery Data

Nonsupervisory Work Groups
(Ngps = 140)

Leadership Variables	Half of Population with higher leadership scores Time 1			Half of Population with lower leadership scores Time 1		
	Means		Diff.	Means		Diff.
	$Time_1$	$Time_3$	t_3-t_1	$Time_1$	$Time_3$	t_3-t_1
Supervisory Support	4.49	4.29	-.20*	3.17	3.72	.55**
Supervisory Goal Emphasis	4.29	4.08	-.21*	3.18	3.69	.51**
Supervisory Work Facilitation	3.84	3.71	-.13	2.58	3.13	.55**
Supervisory Interaction Facilitation	3.93	3.97	.04	2.74	3.86	1.12**
Peer Support	4.21	4.03	-.18*	3.46	3.80	.34*
Peer Goal Emphasis	3.82	3.67	-.15	3.04	3.36	.32*
Peer Work Facilitation	3.63	3.66	.03	2.79	3.09	.30*
Peer Interaction Facilitation	3.47	3.55	.08	2.40	2.89	.49*

*p < .05.
**p < .01.

measures. Under the present plan, the direct effect of the three technological variables on questionnaire measures cannot be assessed. The element in predictions 2b to 3b not assessed with the mean score differences presented above is the indirect effect—the implicit causal chains, associated with technology, operating over time within the social system of supervisory leadership, peer leadership, and group process, and within the psychological system of behavior and attitudes. Partial assessment of these linkages, it is felt, can be obtained through examination of the strength of influence of early social measures on subsequent ones, using cross-lag analysis, while controlling for high and low technology. If, for example, we find that the implied causal nets exist within the social variables for high technology while not manifest for low technology, then the results will aid in indirectly supporting the predictions advanced in Chapter II.

Method of testing causal priorities among the questionnaire measures. The method used to assess strength of causal networks is a simplified modification of the cross-lag correlation technique (e.g., Pelz and Andrews, 1964) where variables measured over time are examined two at a time and a comparison of the sizes of the cross-lag correlations is made. The following illustration presents an example of a typical cross-lag set.

An Example of Cross-Lag Correlations

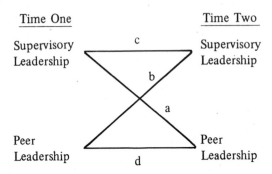

For this example, supervisory leadership and peer leadership are measured time one and time two. The relationship, supervisory leadership$_1$-peer leadership$_2$ (relationship a) is compared with peer leadership$_1$-supervisory leadership$_2$ (relationship b). If the former relationship (a), is greater than the latter one (b), supervisory leadership is said to be a more causal influence on peer leadership, time one-time two, than vice versa. If both relationships (a and b) are equal, then it is said that both variables have causal influence on each other. The cross-lag correlations are also compared to the time one-time two reliability coefficients (c and d) in the following way: If the relationship between supervisory leadership$_1$ and supervisory leadership$_2$ (c) is large in an absolute sense, supervisory leadership$_1$ is said to have causal influence on supervisory leadership$_2$ even when the cross-lag comparisons have established that peer leadership$_1$ is more a cause of supervisory leadership$_2$ than vice versa. If, on the other hand, the reliability coefficient (e.g., c) is not too large, but is as large or larger than the relevant cross-lag coefficient (b in this case), where it is established that the cross-lag coefficient in question (b) is larger than its mate (a), then the reliability coefficient (c) is said to establish additional causal influence on the variable in question.

Since in the present case only general notions of causality are desired, average correlations among the major variables are used. This method of analysis has the advantage of much simplifying an otherwise complex array of data, while having the disadvantages of lower precision and the inability to legitimately utilize tests of statistical significance. The advantage of clarity seems to outweigh the disadvantages on both counts. First, we are interested in the general causal effects of technology on supervisory and peer leadership and group process, of supervision and peer leadership on group process, and the general effects of peer leadership and group process on satisfaction with the work group. Averages among specific items allow these examinations of relationships quite well. Second, since we are dealing with total populations (or nonrandom samples of particular populations) in this study, statistical tests may be less relevant than they would be were we using randomly selected samples of total industrial populations.

The data. Figures 4 and 5 present the results of the cross-lag causal analysis for the insurance company and refinery respectively. These figures show the dominant causal chains over the three time periods for the high-technology group and the low-technology group separately. In order to simplify the figures, cases of reciprocal causality (a causes b, and b causes a) are not shown in favor of presenting only recursive, or more intransive causal chains. The full cross-lag figures of average intercorrelations by site and technology category from which the cross-lag analyses were made are found in Appendix B.

> *Prediction 2b.* Work group leadership, time three, should be found more causally dependent upon initial technology and work group leadership time two, than it will to prior supervisory leadership.

The data in Figure 4 (the insurance company cross-lag analysis) provide some support for prediction 2b. The very stable data in the high-technology group (high coefficients within variables across time) time two-time three might be what we would expect to find between both time one-time two, and

Figure 4

RESULTS OF CROSS-LAG CAUSAL ANALYSIS

Insurance Company Test Site Data

High Technology (N groups = 29)

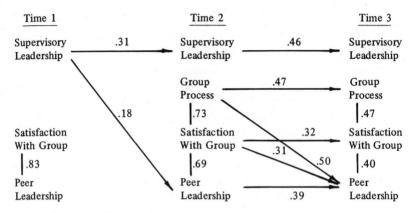

Low Technology (N groups = 22)

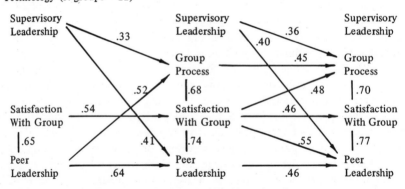

time two–time three, in light of the fact that the high group declined less than the low group over time. The reasoning here is that, given a negative external force, i.e., one working against the direction of sophisticated technology and planned change efforts, technological effects counter to this force would manifest themselves in consistent, stable relationships over time. This is only evident, between time two–time three for the high group rather than for both time lag periods. It is true, however, that peer leadership and group process remain independent of supervisory leadership, which may reflect the constraints of technology. The low group, on the other hand, shows a very strong

Figure 5

RESULTS OF CROSS-LAG CAUSAL ANALYSIS

Refinery Data

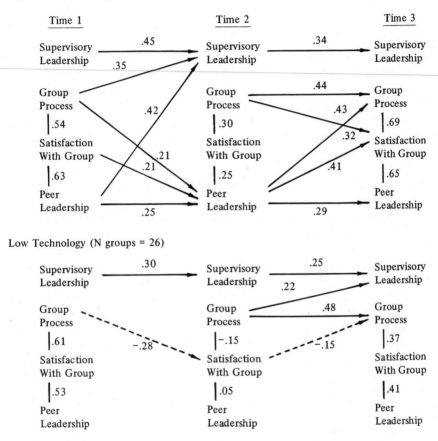

network of causal relationships to subsequent measures of peer leadership and group process. Prior levels of supervisory leadership clearly lead to subsequent levels of group process and peer leadership.[11]

---

[11]This difference in results for the high- and low-technology groups highlights what may be considered an additional connectedness component to the within time and across time components discussed earlier. It would appear that the across time component can be broken down into "stability" coefficients—connectedness within variables across time—and "interrelatedness" coefficients—connectedness among variables across time. That the stability component is different from the interrelatedness component

(continued to page 80)

Figure 5, the lag analysis for the refinery, reveals that causal priorities over time for the high-technology group cannot be assigned to supervisory leadership. In fact, group process and peer leadership time one are seen to have more influence on supervisory leadership time two than vice versa. Since such effects are not evidenced for the low-technology group, it would seem that technology is having a stronger effect overall than is supervisory leadership. It was originally presumed that because the change agents in the planned change efforts would be working primarily on supervision, this variable would subsequently move the others. The refinery data clearly present an alternative system—that of the technological effects on time one (pre-change effort) group process and perhaps on peer leadership, leading to a situation where these variables worked as precursors to supervisory leadership time two. Throughout the time periods measured, supervisory leadership seems not to be a potent cause of subsequent levels of other variables in the refinery data. It may also be true, however, that the time lag periods may have been too long to reveal supervisory leadership effects on subsequent variables.

One assumption behind prediction 2b is that supervision in the high-technology group ought to be a stronger mediating force between time one and time two with regard to peer leadership and group process than it is between time two and time three. Since the time elapsed between time one and time two measurement is quite long, we might not reasonably expect to find a mediating effect of supervisory leadership on peer leadership for that period which could be compared with less effect between time two-time three. In the insurance company (Figure 4), where the lag time is one year between measures, we find a slight effect of supervision on peer leadership t_1-t_2, which disappears between time two and time three as expected. In the refinery (Figure 5), where the time lag between measurement was six months, no supervisory effect is shown for the first period; in fact, peer leadership and group process seem to be the mediators in this case. The difference between time lags in the two companies would lead us to expect that the shorter the time, the more the supervisory effect on peer leadership, but this was not the

---

(continued from page 79)

seems to be evidenced in the insurance company cross-lag analysis described above. Stability, or a maintenance of the *status quo*, as a function of situational constraints on behavior, in face of negative external forces makes sense in the present instance and also helps to explain a puzzling finding described earlier. It was noted in discussion of Table 11 (the connectedness coefficients for the insurance company using combined technology high-low split) that connectedness as a function of level of education seemed a reasonable explanation of the insurance company connectedness results with the exception of the increased connectedness coefficient for the high technology group t_2-t_3. In light of the present data in Figure 4, this result appears to make more sense. If stability coefficients are not related to education as interrelatedness coefficients are, as the discussion of Converse (1964) seems to suggest, then the notion that high connectedness t_2-t_3 for the high-technology group should not be expected to be related to education is born out.

case. It is also true in this case that there are marked differences between the two companies both in technology and in the current organizational environment. In both the insurance company and the refinery, prediction 2b is provided some partial support by the consistent finding that between t_2-t_3, supervisory leadership in the high-technology group is not an unequivocal cause of peer leadership or group process, and that peer leadership time two seems a relatively potent causal force on other variables. In the insurance company low-technology group, supervisory leadership is a strong mediator for both time lag periods as we might expect when technology's situational constraints do not act to make peer leadership more potent and self-maintaining.

One implication, not directly testable in the present case, is that the degree of relationship between peer leadership time three and technological sophistication, although expected to be significant, would be less than that manifest at time two. The indirect effects here, however, can be partially tested by examining causal networks toward peer leadership time two and time three, for high- and low-technology. We predict that we might find peer leadership time two a more ready recipient of causal forces than at time three. This assumes that technology operates to allow improvement in peer leadership t_2 to a certain level, at which point it becomes more self-sustaining and more a causal force on subsequent variables. In the insurance company high-technology condition (Figure 4), peer leadership does become more self-maintaining by time two, and is influenced strongly only by itself, and by group process time two, by time three. For the insurance company low-technology condition, peer leadership t_3 remains strongly influenced by supervisory leadership t_2. In the refinery's high-technology group (Figure 5), peer leadership t_2, although originally self-maintaining, is a recipient of influence and becomes an influence force on time three variables. The refinery low-technology group, on the other hand, shows no such pattern; peer leadership remains uninfluenced and noninfluencing throughout. Thus, tentative support is given this notion in both insurance company and refinery.

Insofar as it seems that the results cited establish that peer leadership becomes more self-sustaining over time in high-technology groups, the reasons advanced in Chapter II (justifying predictions 3a and 3b) to help explain this maintenance remain valid for test. These notions involved the reinforcing effects of more favorable attitudes toward the work group in the high-technology condition, based on the goodness of fit between technological constraints, and new behaviors (the outcome of the planned change program), upon subsequent behaviors.

Prediction 3a. Satisfaction with the work group, time three, is expected to be causally dependent upon work group leadership, and autonomous group process time two.

The cross-lag analysis, as an evaluation of causal priorities and method of examining indirect technological effects, provides a way of testing the assumptions behind this prediction. The assumptions behind the prediction involve Festinger's notion of cognitive dissonance (Festinger, 1957), and were described above in Chapter II as follows:

> ... having experienced the better fit between the new ways of behaving and the technological situation, group members are faced with a dilemma—their favorable and early socialized values and attitudes toward more traditional management styles are in conflict with the favorable experience of group leadership at time two. The constraints of the situation have not, and do not lend themselves to reversion to the behaviors consonant with the older values and attitudes. Thus, new attitudes, consonant with the new behaviors are likely to be established between time two and time three measurement as a function of the reduction of cognitive dissonance (pp. 33-34).

Given this line of reasoning, we may look at the causal nets among group process, peer leadership, and satisfaction with the work group for behavioral changes leading to attitudinal changes. Since the insurance company represents a special case of maintenance of original behaviors in face of negative external change forces, it will be discussed following an examination of the refinery data. In the refinery high-technology group (Figure 5), satisfaction time one has causal influence on peer leadership time two, and peer leadership and group process time two have causal influence on satisfaction time three. This is quite in line with the results expected—the changed behaviors lead to more favorable attitudes, but only after a period of time (the simultaneous relationships at time two being quite low compared with those at time one, or time three). For the refinery low-technology group, on the other hand, the results are dramatically the reverse. Group process time one has an *inverse* causal influence on satisfaction with the work group time two, and this in turn has an inverse causal influence on group process time three. Lower group process time one produces higher satisfaction which, in turn, has a slight perseverative effect at time three. It is also clear for the low group, however, that the simultaneous relationship between group process and satisfaction is positive, which suggests that Festinger's notions of cognitive dissonance are operating at this level as well—albeit slower than that shown for the high-technology group. Thus, we may conclude that changes in group behavior tend to maintain themselves via changes toward more consonant attitudes over time. The insurance company data (Figure 4) do not show this effect; in fact, original attitudes seem to be maintaining subsequent behaviors. For the high-technology group, the simultaneous relationships at time three show a marked decline from those at time two, which suggests that the more peer leadership and group process declines, the less linkage there is between good performance relative to previous periods, and good feelings about it. Knowing what we know about the rates of decline in group process and peer leadership (Table 13), it seems clear from the present causal analysis that both high- and low-technology groups in the insurance company are maintaining what they can of prior behaviors as a function of resistance to attitude change. The situational constraints of technology in the high-technology group provide more counter-force against negative change than those constraints for the low-technology group. Therefore, they provide more successful resistance to the negative forces than in the low group, given similar attitude-to-behavior causal nets.

Prediction 3b. The relationships between time two satisfaction with work group, and time two peer leadership and group process are expected to be lower than those between their time one and time three counterparts where the technology is sophisticated.

Prediction 3b was advanced in an attempt to assess the potential for increased permanence of change in the high-technology group given high inter-relationships between behaviors and attitudes at the final measure. Part of the data relevant to this prediction were described above with regard to Prediction 3a. The refinery high-technology data (Figure 5) clearly support this prediction. It was also noted for the low-technology group that, although it was much weaker, the same effect seems to hold there as well. In the insurance company, as mentioned above, the data are reversed with low-technology group being higher than the high-technology group in these relationships and the relationships for the high-technology group dropping in size from time two to time three. Since it is assumed that the high-technology group has more constraints against the external negative change force, the lower inter-relationships between behavior and attitudes may suggest that there is less permanence in the negative change for this group by the third measurement period than for the low group.

Summary. This final section of Chapter IV has shown that where test was possible of the specific predictions proposed in Chapter II, the results form a mixed picture. Given long time periods between measurement, the expected relationships and causal nets among the measures were not always in evidence; it is unknown whether these could be obtained with shorter measurement periods. The predictions made regarding the effects of behaviors on attitudes, and of these attitudes on subsequent behaviors, were supported quite well.

V

DISCUSSION OF RESULTS

The purpose of this chapter is to discuss the meaning of the results presented in Chapter IV. This discussion will deal exclusively with the test results of Chapter IV, since the meaning of the preliminary analyses in Chapter IV were assessed as they were presented in order to justify continuing with the test analysis. The present discussion will be guided by the order in which the main results were presented in Chapter IV.

It is unfortunate for both the author and the reader that there is such a paucity of analytic technique for dealing with interaction effects and their results in nonexperimental research. It is unfortunate for the author on at least two counts. First, the elegance of design and implementation in causal analysis using additive models leads to clearly definable results—a thing pretty much is, or it isn't—the outcome is straightforward. With interaction, on the other hand, nothing is straightforward and main effects are difficult either to tease out or to support. Second, interactional data are extremely difficult to describe and discuss. Interactional results seem, at least to the author, to require that specific findings are described in several ways since none seems to do an adequate job alone and the description of the mass of data leads to a thoroughly confusing series of gigantic tables which go beyond human perceptual abilities to absorb as an entity. Easy organization of such data is not possible, and summary organization of the data obviates all but the most rudimentary statistical treatment. It is unfortunate for the reader, because it is difficult for the author. Anything so difficult to write about must necessarily be difficult to read about. As Chapter IV revealed, I have chosen to deal with the data in summary form where possible. This course, it is hoped, has cut a path through what seems otherwise a bewildering array of findings. Since much control in the data is impossible using this method, this study must, in retrospect, be considered even more exploratory than was originally intended.

Connectedness

This idea seems to strike a resonant chord with the open system notions of relatedness and interdependence of organizational subsystems described by Katz and Kahn (1966), and the system interdependence versus functional

85

autonomy of systems described by Gouldner (1959). While these authors usually deal with larger subsystems than those in question here, it seems quite reasonable, and Gouldner states explicitly (pp. 421-422), that these notions operate equally well in smaller subsystems such as the intrasocial-system elements we are using in the present case—supervisory leadership, peer leadership, and group process. Tables 7-9, and 11-12 presented results in a form of coefficients of connectedness. The data in Tables 7-9 revealed that each of the three technological variables was having observable effects on the patterning and strength of relationships among the dependent variables. This outcome was enhanced by use of a combined measure of the three technological variables (Tables 11-12). In total, these examinations also revealed that connectedness is probably related to education, following the finding by Converse (1964) that more education is associated with a higher degree of consistency and interrelatedness of attitudes and behaviors. It seems that there are three components of connectedness measures in the present study: a) *within time connectedness* (probably a poor measure because of methodological problems surrounding relationships among variables measured at the same time from the same source), b) *stability*, measured by intravariable relationships across time, and c) *interrelatedness*, measured by intervariable relationships across time.[12] Of these three components, within time connectedness is related to education or technology; stability seems more related to technology than to education, and interrelatedness seems more or less equally related to technological sophistication and to level of education.

Connectedness and change efforts. Comparisons between Tables 7 and 8 indicated that connectedness was associated with the planned change efforts. Table 7 and Figure 4, the insurance company test site data, showed immediate differences in interrelatedness t_1-t_2 for the high- and low-technology groups which matched the educational levels of the groups and variables, and which coincided with the onset of planned change efforts. Table 8, the insurance company control site data, showed no t_1-t_2 connectedness differences between the high- and low-technology groups, but showed the same pattern of t_2-t_3 connectedness for the three technological variables between the groups at the onset of the planned change effort as that of the test site t_1-t_2 connectedness. This exact correspondence of connectedness patterns among the three technological variables, high and low, following introduction of change efforts for the test and control sites, is interpreted as evidence of change in connectedness as a function of planned change efforts, educational

---

[12]This third component may, in turn, be broken down into two components: general and equivalent interrelationships across time, and interdependence (where causal estimates are determined to exist, and given subsequent measures can be said to have dependence on certain prior ones). That the former component is more like the stability component with regard to change, and the latter component is the one which is related to change seems likely.

level, or technological sophistication or, more likely, a combination of all three. Although it might be that the pattern in relation to the change effort is, in fact, a chance finding, the number of controls employed (e.g., same measures, same time periods, same company, sites matched originally for performance effectiveness, and an active intent not to introduce change in the control sites until after the second measure) would seem to rule out this explanation. In fact, considering as a usual response in such natural experiments the unwillingness of control groups to accept a condition of not receiving what is thought to be valuable and helpful treatments, it is encouraging to note the results reported above.

Combined technology classifications and connectedness. Tables 11 and 12 presented data which showed that finer discriminations of sophistication of technology enhanced and increased the extant differences in connectedness patterns noted in Tables 7 and 9. This is taken as further evidence that technology influences connectedness.

For the refinery data, it was shown that because the combined technology high-low groups did not differ with regard to education, the differences between these groups in connectedness patterns in Table 12 were most likely due to technology. For the insurance company, on the other hand, the combined technology measure was found to be associated with education (low-technology group had higher education) which may help to explain the high connectedness scores for the low group, but do little to explain the high connectedness score t_2-t_3 for the high group.

The differences in connectedness patterns between the insurance company and the refinery present something of a problem. We might expect that high technology should affect connectedness in the same way regardless of company. Since differences in educational level were manifest for the insurance company high-low technology groups, and not for the refinery, the effect of education may in part explain this difference. As another possible reason for this difference between companies, we must recall that prior research, reviewed in Chapter I, suggested that real differences in the effects of sophisticated technology *could* exist between white-collar and blue-collar systems, even though this was not established.[13]

Another reason why the connectedness patterns between high- and low-technology groups may differ between the two companies might be a function of the reaction of groups to negative change forces, a reaction conditioned by technology. It was noted in Chapter IV that the overall insurance company

---

[13]With regard to this difference between organizations, it is interesting to note that the mean scores for the three individual technology variables within the high and low combinations for the two companies, shown in Table 10, reveal big differences only between the low groups and only for the input and throughput variables. Since the refinery was expected to have much higher technology scores than the insurance company, this result is puzzling.

survey results were known to decline (shift away from participative management) over time, and that the results presented in Table 13, the mean score changes controlled for technology, similarily declined. (The reasons for the decline, as mentioned earlier, were never unequivocally determined in the overall study, but were considered to be the effects of a top management group which turned to a position of more authoritarian management during the study as a result of severe market pressures from the organization's external environment.) We have seen in Table 13 and Figure 4 that the rate of decline in mean score responses on the dependent variables was slower for the high-technology group (with lower overall connectedness, but somewhat greater stability) than for the low-technology group (with higher overall connectedness). These negative changes are changes which are opposite in direction to those we assume are, at least in part, predicated by sophisticated technology. This finding provides us with an elaborative statement for our basic hypothesis: *sophisticated technology, it seems, not only will facilitate change efforts which are in a direction consonant with that determined by the technology, but sophisticated technology will aid in resisting change efforts which are in a direction opposed to that determined by the technology.*

In the refinery, connectedness was greater for the high-technology group and greater positive change was exhibited for the group as well. In the insurance company, connectedness was greater for the low-technology group, and greater negative change was exhibited for the group as well. Since connectedness coefficients were greater in the main for the insurance company low-technology group, and since the degree of mean change over time was greater for that group than for the other, the statement that high connectedness is associated with greater organizational change *in any direction* holds for the data presented in Chapter IV. Although connectedness coefficients increased in strength for the insurance company high-technology group for time two–time three, Figure 4 revealed that this was accounted for by an increased intravariable stability rather than by an increased intervariable interrelatedness. It is concluded here that the patterns of connectedness for high- and low-technology groups probably differ between the insurance company and the refinery because of differences in educational level in the insurance company and differing directions of change forces in the two companies, coupled with the forces generated by sophisticated technology, being either with or against those forces. It is assumed that, were the insurance company operating with the same management climate and market conditions as the refinery at the time of the study, the across time connectedness patterns between the two companies might have been more similar.

Given the relationship between connectedness and change, and taking the finding from Mueller's (1969) study that the more highly educated are more adaptable and accepting of job conditions and the Converse (1964) finding that education is associated with interrelated values and behaviors, we might conclude that the differences in rate of dependent variable mean score change between the insurance company high- and low-technology groups (Table 13) could be explained by the effects of education alone. This would question the validity of the elaborative statement made above that sophisticated technology can facilitate some change forces and attenuate others. We know, for example, that the insurance company low-technology group had higher levels of education than the high group. It had more connectedness, and it changed more away from participative management, all reasons in favor of the education-as-sole-cause thesis. But we also know, upon examination of Figure 4, that the one instance of high across-time connectedness for the high-technology group took the form of stability rather than interdependence (or the existence of separable causal priorities). We assume that the higher education levels in the low-technology group acted to increase the interrelatedness (interdependence among measured variables), and aided member acceptance of change away from participative management. We are also assuming, however, that the sophisticated technology's situational constraints in the direction of participative management acted to *enhance the stability* within major variables over time, thus enhancing resistance of that group to negative change forces (forces in a direction away from participative management).

The preceding paragraphs attempted to explain that the differences in connectedness patterns between the two companies may not be reflections of differences between white-collar and blue-collar technologies. These reasons do little to explain why the high-technology group in the insurance company was initially so much lower than its low-technology counterpart in absolute mean levels of the dependent variables (Table 13). It was hypothesized in Chapter II that if sophisticated technology could operate alone in the direction of more positive peer leadership and group process variables, then initial levels of the variables would be higher for the high-technology group than for the low-technology group. The difference from this in the insurance company seems to be a product of education. Whether educational differences, however, are a function of differences between blue- and white-collar technologies themselves, or differences in management (or hiring) practices vis-a-vis technology in blue- and white-collar organizations, it is impossible to say. Regarding this, it will be recalled that Mueller's finding that sophisticated technology is related to higher levels of education was in no way confirmed in the present study. Chapter I described the research data on white-collar technology and worker behavior as limited to cases where poor management practices (rather than qualitative differences in effects of technology in white-collar organizations) might have accounted for worker behavior and attitudes. There is no evidence in the present study to clarify this point—all that seems reasonable to say is that there may be differences in worker response to sophisticated technology (regardless of education) between white-collar and blue-collar organizations, but these differences are probably not measured in the present study.

Technology, connectedness, and change. Connectedness is associated with education. Connectedness is associated with sophistication of technology. Sophisticated technology is associated with organizational change. Connectedness is associated with organizational change. How do these statements link together with regard to the modified global postulate that technology facilitates a particular kind of planned change effort, and obstructs another particular kind?

The basic mediating concept used in Chapters I and II to explain effects of technology on reaction to change was that of inexorable situational constraints acting on worker behavior in the same direction, and parallel to that of the urgings and exhortations of the change agents. There was no measure of these situational constraints; they were presumed to exist, with vector strength varying as a function of technological sophistication. The refinery data in Chapter IV clearly showed that the strength of results of the planned change effort were directly related to technological sophistication as predicted. Additional data in Chapter IV revealed that technological sophistication was also related to connectedness (or system strength, an overall interrelatedness among the dependent variables) over the time planned change attempts were effected. The interrelatedness component of this connectedness appeared related to the degree of change, not necessarily in the direction expected in light of the planned change efforts, but change in general in the dependent variables. The idea that interrelatedness is a facilitating condition for change is not original in this study. Katz and Kahn (1966), as cited in Chapter I, assert that external forces lead to pervasive organizational change because of an interrelationship among organizational subunits. This "interlocked set of gears," or "row of dominoes" sort of notion regarding pervasive organizational change is a basic tenent in open system theory. Gouldner (1959) also makes the same point coming from the opposite direction—that systems having a low degree of system interdependence (or high functional autonomy) will be more likely to survive external forces (i.e., resist changes better) than systems with high interdependence. Thus, it seems reasonable to extrapolate from the statement that "interrelationship among substructures leads to organizational change" to the statement that "the greater this interrelationship, the greater the change given the same strength of the external change vector." This is what the data in Chapter IV seem to force us to conclude. Although the measures of connectedness components are quite crude, the effects seem strong and unambiguous. Since no research appears to have been done using interrelationships among social subunits as a variable, there is no prior evidence to either confirm or deny these results. They must stand as a single, serendipitous finding at this time. Another, and less understood finding is, as was noted above, that changes in the degree of connectedness appear associated with planned social change efforts. Aside from the concept of successful change entre as "unfreezing," proposed by Kurt Lewin, there is little in the literature to anticipate this result. Gouldner, for example, specifies several sources of low interdependence or functional autonomy, all of which are structural or technological in nature.

Results in Chapter IV suggest that technology acts to condition the relationship between change efforts and change results in at least two ways. *First*, technological sophistication, through its situational constraints on specific worker and supervisor behaviors, acts to influence the efficacy of change efforts. If those efforts are in the same direction as the constraints of sophisticated technology (i.e., greater group autonomy, participation in decision-making, responsibility), the efforts are facilitated because people are being asked to behave in ways more consonant with the constraints of the technological system. If those efforts are in a direction away from more participative management, the change efforts are attenuated by sophisticated technology because people in these settings are being asked to behave in ways which are difficult to achieve because they run counter to the technological constraints. *Second*, technological sophistication seems to condition the relationship between positive change efforts and effects by increasing system strength or interdependence among social subsystems. It has been noted that both technology and education are similarly related to across time intervariable dependence which, in turn, is related to change results, but that education and technology are not necessarily related to each other. The second effect seems different from the first one, described above, in that it (the former) describes a more molar, systemic effect, while the latter describes a molecular or individual behavioral effect.

With reference to the second effect, it seems not unreasonable to assume that technology, being a systemic variable itself, would have a systemic effect as well as individual behavioral effects on the members of technologically-oriented work teams as originally postulated. It will be recalled that the idea of situational constraints on individual behaviors was shown to be evidenced in the prior literature on technological effects, while more systemic effects were less well documented. The results in the present study, insofar as they provide some quantitative evidence of systemic effects, seem to lend support to the position of the socio-technical system theorists that technical and social systems interact together as total entities.

A Sampling Problem in the Combined Technology Measure

At the time the technology evaluations were combined, using both the optimization, and maximization procedures, the composition of groups in the high-technology and low-technology categories was examined. It was found that the composition of groups when using the optimization procedure differed from the composition obtained when using the maximization procedure. In addition, the optimized high and low categories contain groups with seemingly nonrandom similarities in factors, aside from education, not originally included in the test of the main effects. In the insurance company, nearly all the high groups obtained using the optimization method were groups of female employees. This could have been a problem, as we have seen that sex of respondent had some effect on the dependent variables (Table 4). Using the optimization method in the refinery, all the high-technology groups were in the maintenance division, while all the low-technology groups were in the administrative-engineering division. This outcome in the refinery was a potential problem since all the groups in the high-technology category received planned change inputs from one in-plant change agent, while all the groups in the low category received inputs from another change agent. In the case of the insurance company, we might conclude that the results obtained actually reflected differences between women and men employees, while in the case of the refinery, we might say that the results obtained were those of differences in change agents. In the discussion to follow, these concerns will be shown less important than they may appear.

Insurance company. Using optimization, the combined technology high category contained mostly all female groups situated low-middle in the organization's hierarchy, and the low category contained groups composed of all men, or a combination of men and women. These low-technology groups were also located in the low-middle of the hierarchy. The maximization high-technology category contained groups with all women, and groups with men and women, low to low-middle in the hierarchy.[14] The maximization low-technology category contained mostly staff groups composed of men, situated in the middle of the company hierarchy. For the purposes of our concern (the optimized high category being composed of all women groups), a comparison of results obtained using the maximization procedure, and the results using the optimization procedure, will in part provide reasons to reduce that concern.

Table 10, it will be recalled, compared the mean levels of the three specific technological indices for the combined technology groups derived using both the optimization and maximization procedures. The results in Table 10 revealed that using either method produced the same technological scores for the high groups—that is, the combined technology high category represented the same average technological scores on the three variables (input, throughput, output) regardless of whether the groups were chosen by optimizing all three variables, or maximizing the first two. Since the sex of

---

[14]It should be noted here that the problem of sophisticated white-collar technology having a great number of repetitive, fractionated, transitional tasks as described in Chapter I is not relevant in this sample. The insurance company had few key punch groups, most of these tasks being handled within groups established on product, not functional, lines. In neither optimization nor maximization were pure key punch groups selected in the high or low categories.

respondent mix differed using the optimization or maximization procedure, it would seem that between the two of them, the maximization high-technology category would best be used to eliminate the concern of the effect of respondent sex on the subsequent results. Tables 17 and 18, in Appendix C provide the mean score change data using the maximization procedure for categorizing high and low technology for the insurance company and the refinery respectively. These tables parallel Tables 13 and 14 in Chapter IV except for the method by which the high- and low-technology categories were obtained. Comparing the mean scores for the high-technology group in Table 17 with those in Table 13, it is quite clear that for both time one and time three periods, the maximization high group scores the same as the optimization high group. That the insurance company high-technology group used in the test analyses was composed of all female groups does not seem to be of vital concern. Using the maximization method for selecting the low-technology group, however, does produce different results from those obtained using optimization. Results in Table 10 revealed that although the maximization procedure obtained significantly lower scores on input and throughput for the low-technology group, it also produced a significantly higher score on output control. The low group mean scores in Table 17 (Appendix C) are the same as those in Table 13 (Chapter IV) at time one, but the time three measures are higher in Table 17, making the rates of decline in mean scores in Table 17 for the high and low groups similar to one another. This suggests that moderately highly placed, low-technological staff groups resisted negative change about as well as did groups low in the hierarchy with higher technology (either men and women, or women alone), and better than did groups lower in the hierarchy with low technology. This is not surprising since most higher status people, with more freedom and more flexible role boundaries, are in a better position to resist change than lower status people. What is surprising is that they could resist no better than low status groups with sophisticated production technology.

Since we have found education to be a potential confounding variable in the model proposed, it seems of interest to describe the effects on this variable in the insurance company data using the maximization procedure. For the maximization high-technology group, the education mean = 4.28, exactly the same mean obtained using the optimization method. For the maximization low-technology group, the education mean = 5.08, higher than that obtained using the optimization method, but not significantly so. This maximization low-technology education mean is, however, greater than its maximization high-technology counterpart to a highly significant degree (t = 5.21, p < .01). It is clear from these data that the maximization method could not have been used to control for the effects of education in the insurance company data.

Refinery. Using optimization, the combined high-technology category contained nothing but maintenance department groups, while the low category contained only groups from the administrative-engineering department. It will be recalled from Chapter III that the refinery contained three departments: production, maintenance, and administrative-technical. Each of these three departments received help and guidance in the planned change program by a different in-plant change agent. By that token, it seems not unlikely that the results obtained using the combined technology high and low categories by optimization could reflect the effects of difference in competence between change agents rather than technological effects. This concern was enhanced by the result that the high-technology group obtained by optimization was from maintenance rather than manufacturing, since it had been originally expected that the latter group was more technologically sophisticated than the former and should have scored higher.

Using the maximization procedure for obtaining combined technology scores, the composition of groups in the refinery high-technology category was a rather even mix of groups from the manufacturing and maintenance department.[15] The mean levels of the three technological indices for the high-technology category using the maximization procedure, however, as shown in Table 10, were less than those using optimization—both input and output levels were lower using maximization, even though throughput was much higher. This suggests that even though the manufacturing department has more sophisticated machines, its employees must spend more time setting up for unstandardized raw materials, and have fewer extra-supervisory feedback channels. Comparing the mean score change results for the high-technology categories produced by both methods (Table 14, Chapter IV, and Table 18, Appendix C) reveals that although somewhat attenuated, the change results using maximization are similar to those obtained using optimization. Specifically, Table 18 shows peer interaction facilitation and group process initially higher for the high-technology category, and shows peer interaction facilitation, work facilitation, and supervisory work facilitation improving more for the high group over time. Improvement for supervisory and peer support is also shown for the high-technology group as well.

The preceding results were discussed with the university change agent who was responsible for coordination of the refinery study, and the change efforts expended there. It was his opinion that the differences obtained between maximization and optimization procedures, and those obtained between high and low categories using optimization were not due to competence differences among the in-plant change agents. In fact, it was his belief that the change agents in the maintenance and administrative-technical departments were equally competent, although they differed some in personal style.

---

[15]The low-technology category, it will be recalled, had the same composition of groups using either method of combining technological scores.

The education mean scores for the refinery high- and low-technology groups using the maximization method showed even less difference between them than did those using the optimization method, the difference for which was found to be nonsignificant. For the maximization high-technology group, the education mean score was 4.18, while for the low group it was 4.15.

The finding that the optimization procedure produced stronger questionnaire mean score differences over time in both companies suggests that the facilitating effects of sophisticated technology on planned change efforts are best shown when all three technological variables (input, throughput, and output-control) are taken into account. It provides a basis for claiming that, even in industrial organizations, technological classification schemes should include more than mere assessment of production technology hardware. The results in the refinery are also interesting in that they show greater changes toward participative management among high-technology, blue-collar workers (maintenance department) than among white-collar, professional employees (administrative-technical department) with less sophisticated technology. We might reasonably expect that white-collar, semiprofessional employees, with their assumed greater identification with the company, would accept the sort of changes we are dealing with here more readily than the blue-collar workers. The reverse findings, when controlling for technological sophistication, are interesting in this context although not definitive. Lack of design strength obviates the possibility of really testing this effect, since attempting to control for white-collar, blue-collar characteristics while controlling simultaneously for technological sophistication would reduce the n's within cells to a size too small to be usable.

The Causal Analysis and the Specific Hypotheses

As noted in Chapter IV, many of the specific predictions dealing with assessment of strength of causal influence by technology could not be directly tested. The implication of many of these predictions were examined, however, for what light the data in Figures 4-5 could lend. The predictions themselves fell into three categories: influence of early supervisory leadership and technological sophistication in planned change, technological sophistication and self-maintenance of peer leadership in planned change, and the reinforcing effects of changed attitudes on subsequent behaviors in peer leadership and group process under sophisticated technology during planned change.

The influence effect of supervisory leadership. Supervisory leadership was expected to play an influence role early in the time sequence on peer and group behaviors. It was assumed that because the change agents were working primarily with the supervisors, and not so much with the groups, that supervisory behavior would change first. Peer leadership and group process could have an effect, but the initial change forces would come from *early changed* supervisory behavior. To a large extent this did not hold in the data. In the high-technology groups, supervision remained independent of the other variables, had mutually influencing effects, or was influenced by peer and group behaviors over the time periods. It may be speculated that the time between the first measure and the second was simply too long to capture this effect which may have taken place in a matter of a few months. It is also not reasonable to expect supervision time one to have effect, being simply a base line for change. We might more realistically expect to find changes in supervision by an unmeasured post-change period, time one-and-one half, to subsequently influence peer and group behavior time two. The other part of the specific predictions, that supervision in the high-technology group would have lower influence on peer and group behavior by time three seems partially supported at least to the degree that supervisory leadership does not influence other variables except in a nonrecursive system where the influence is mutual.

Self-maintenance of peer leadership. The results supporting these predictions were more certain than those used to test the effects of supervisory leadership. It was expected that by time two, peer leadership would have been strengthened to the point where it would maintain itself and influence group process as well. The results for the refinery show this effect very well for the high-technology group, and show the not unexpected absence of effect for the low-technology group. Since, as it turned out, the refinery was the only site of the two for which the change process was, in general, working as it should have, these results are really the only ones to consider as a test of the predictions. The insurance company was not changing behaviors up, but changing down instead. In view of this, self-maintenance of peer leadership in the insurance company takes on a different character—that of providing stability in resistance to change. In fact, between time two and time three, the insurance company high-technology group showed considerable stability, or self-maintenance of all variables measured, while the low group showed little stability of supervisory leadership coupled with strong influence of supervision on the other variables over time. We can speculate that as supervisory leadership in the low-technology group was changing down, it was pulling the peer and group behaviors with it—the overall result being a faster decline than that noted for the high-technology group.

The reinforcing effects of changed attitudes on subsequent behaviors. Once again, the refinery data provide the only real test of the predictions because the change effort was, in general, working there. It will be recalled, it was postulated that permanence in change effects would be enhanced where favorable attitudes (generated by the goodness of fit between new behaviors and the work situation) reinforced simultaneous or subsequent behaviors. We stated that such goodness of fit would be greater in groups with high technology than in groups with low technology because of the technology's situational constraints toward autonomous and responsible work groups. The refinery data clearly support this contention. Peer leadership and group process by time two in the high-technology category are stable but are not yet strongly associated with satisfaction with the group. High peer leadership at time two is associated with high group process time three, and both peer and group behaviors lead to subsequent time three satisfaction which, in turn, is strongly associated with simultaneous peer and group behaviors. The process of unfreezing and refreezing between time one and time three with regard to these measures is clearly shown for the high-technology category. The low-technology category in the refinery shows a slight tendency toward this effect, but it is also coupled with a set of inverse causal relationships. This suggests that the element of situational constraint in sophisticated technology plays a real role in establishing early permanence of favorable change via the reinforcing effects of attitudes on behaviors. This result also lends some additional support for the cognitive dissonance model of attitude change caused by counter-attitudinal behaviors. We have assumed that the cognitive dissonance forces are more potent in the high-technology category because of the situational constraints on behaviors which are already in the direction of the role changes being urged by the change agents. Behavioral changes are greater and more uniform here, following the change efforts because the employees began with a head start over those in the low-technology category. In the low-technology category the change agents were urging role changes which did not match the work situation so well. Employees were slower to change and probably were less consistent and stable in the new role behaviors because they had no rigid situational constraints in the direction of the new behaviors. Role boundaries are flexible, and employees in the low-technology categories could deviate by testing the extent of those boundaries which were their primary constraints. Thus, by the time two measure, groups in the low-technology category found themselves behaving more in line with old values and attitudes than did employees in the high-technology groups. Hence, there was less cognitive dissonance, leading to less attitude change and less subsequent behavioral change and, therefore, less permanence in the new behaviors.

The insurance company data seem pretty clearly to reveal the effects of maintenance of old values and behaviors in the service of resisting negative change. They, therefore, cannot be used to test the reinforcing effects of attitudes in permanence of change.

Some Limitations in the Method, and Implications for Future Research

The results discussed above reveal that, in general, the predictions and expectations of the study were born out rather well. This study was attempted in an effort to develop and study an instrument for measuring sophistication in several factors of technology, and to see how well such a measure could predict to organizational behaviors. It was part of an ongoing longitudinal research project in which some data had already been collected in several sites and was, by that token, necessarily somewhat limited in scope and sample. The present section provides a discussion of some of the research limitations in this study, and some of the activities which seem useful to undertake in additional studies using these methods.

Much of the literature cited in Chapter I dealt with the effects of technological change on worker behavior and attitudes. This literature was useful in estimating effects of sophisticated technology since it provided natural before-after controls. It is also true, however, that the effects of technological *change* might be different from the effects of technology as a static state. Some studies did deal with a static technology (e.g., Blauner, Woodward), but the greatest share of the literature on which the predictions in the present study were based were studies of technological change. The present study would have been a better design, relative to the existent knowledge, if it had been a study of the conditioning effects of technological *change* on planned social intervention. This was impossible, however, given the availability of study sites, none of which was currently undergoing, or had recently undergone technological change of any magnitude. Given the static measure of technology which was used, the predictions extrapolated from the technological change literature worked out rather well.

That the measure of technological sophistication was undertaken in mid-1969 and was used as an indicator of technology before the ISR studies were undertaken one to three years earlier, was also a drawback. Although it is true that the judges were instructed to evaluate group technology as it was at the time the study began, such control via instruction in retrospective data is a weak substitute for coincident measurement.

The results obtained in this study regarding the relationship between education and the technological measure present another potential problem. Regarding the three technological indices, education was shown to have a strong negative association with the input variable and a strong positive association with the output variable for both companies studied. That the evaluations of technology may really reflect judges' evaluation of work group education levels is a possibility. This potential diminishes somewhat when the combined technology classes are examined, however, for the refinery, no association between technological scores and education scores was noted. For the insurance company, on the other hand, rather strong differences in education between the high- and low-combined technology groups were evident. Whether these differences in the insurance company reflect real associations between technology and education levels or simply evaluations of education, it is impossible to say. It must be acknowledged, however, that these results in the insurance company are opposite to those found in Mueller's study of the total labor force where high technology was associated with high education, and which, because of its scope, must be considered a rather definitive study with regard to this relationship. Mueller's measure of technology was an assessment of what we are calling throughput, however, so the results are not directly comparable.

In administering the technological measurement, a limited number of judges were used. The usual case was that each judge would evaluate a separate and unique set of work groups; each group in the sample was evaluated by only one judge. This had the effect of producing discrete rather than continuous distributions of group scores for each measure. This condition was attenuated some with the combining of individual items into mean score technological indices, and would have been further reduced with the combination of all three indices into a mean score total technology measure could it have been justified. In any event, this use of few evaluators produced a problem of inability to obtain reliability scores for any but the pretest site. Such a situation could be relieved by using multiple judges for all groups.

The manner in which the interaction effects found in the present study were treated represents one of the two methods possible. It might have been more useful to determine the functions underlying the interaction effects in the two sites and to transform the dependent variable data in order to obtain a linear function and hence an additive model. In the present instance, this second alternative was rejected since the three technological indices revealed potentially different interaction functions among one another and within each for the various relevant relationships (creating a very complex task of best-fit single correction, or multiple correction for the dependent variables).

The sites used in the present study represented examples of white- and blue-collar industries. This was fortuitous for the purposes of studying differences in technological effects between such industries, since the literature cited in Chapter I suggests that differences in effects might exist. Unfortunately, the two sites were not only different in primary mission (i.e., service vs. manufacturing), but were different in their overall response to the planned change program. That the insurance company was evidently experiencing the effects of some outside force which disrupted the change efforts made it different from the refinery for many comparative purposes. To have matched the refinery with an insurance company or other service industry with a more agreeable management climate or market environment would have allowed us to more categorically state what the effects of white-collar technology were on planned change efforts.

VI

ADVANCED TECHNOLOGY AND WORK GROUP BEHAVIOR IN A SETTING OF PLANNED SOCIAL CHANGE: A REPLICATION STUDY

Introduction

The preceding chapters described findings in support of the hypothesis that the state of automation is associated with a more autonomous work group structure at lower levels in industrial organizations. A test was also made of the hypothesis that advanced technology facilitates social change in the direction of democratic and autonomous group process. Earlier chapters presented a review of the literature, a model of technological effects, a measure of sophistication of technology, and some longitudinal survey data from two organizations in a test of the above hypotheses.

The present chapter describes the results of replication of the design used in the original study in several other organizational sites. In the present chapter, data from the refinery organization, used in the initial study, are compared with similar data obtained in organizations engaged in glass products fabricating, metal products fabricating, and continuous-process plastics production.

This replication differs from the initial study in two major respects. First, the initial study sites undertook at least three longitudinal survey measurements, while the replicate sites have only been surveyed, at most, twice. Second, the initial study sites were quite large, with over 100 nonsupervisory work groups. The replicate sites include, respectively: 20 groups in two measurement waves; 40 groups in two measurement waves; and 140 groups in one wave. These data, therefore, lend themselves most suitably to the test of a static assumption that technology is associated with autonomous group process and less suitably to a test of the facilitating effects of technology on the social system change.

The Study

The purpose of the overall study was to test several notions advanced by organizational theorists regarding the effects of sophisticated technology, or automation, on job-related behaviors in work groups. Among these were two hypotheses: *first*, that sophisticated technology, in and of itself, is associated with more autonomous and participative group process; *second*, that sophisticated technology will facilitate planned change efforts directed toward increasing participative group process.

Methodology

This chapter reports data obtained from four industrial organizations. First, the present chapter utilizes results obtained from additional analyses undertaken with refinery data reported in Chapter IV. These data were described as obtained from the responses of over 1,000 persons in 140 nonsupervisory work groups employed by a large petroleum refinery. Respondents completed questionnaires dealing with job-related matters. These questionnaires were completed by respondents on three separate occasions over a period of 12 months.

The second site used provided data from 123 persons in 22 nonsupervisory work groups in a glass products factory. Respondents completed the same questionnaire used in the refinery. These questionnaires were completed by respondents on two occasions in a period of 12 months.

In the third site, a continuous-process plastics producer, 322 persons in 52 nonsupervisory work groups completed questionnaires similar to those used in the refinery and glass products factory. Once again, the questionnaire was completed twice in a period of 12 months.

The fourth site, a metal products manufacturer, is still in the initial stage of measurement with only one questionnaire survey. Data here include responses from some 950 persons in 148 nonsupervisory work groups.

A planned change program toward more participative management was introduced following the initial survey in each of the four sites.

Finally, the judgments of some in-plant people at each site were used to obtain evaluations of the sophistication of work group production technology for groups within that site.

The analytic design involved controlling for sophistication of technology and examination of the survey results of group responses to questionnaire variables measuring participative leadership, group behavior, and satisfaction.

Independent variables. The measure of sophistication of technology was considered both an independent variable and a conditioning variable. As an independent variable, the effects of technology on pre-change levels of leadership, group process, and satisfaction were examined. As a conditioning variable, the effects of technology on rate of change in these variables, and where possible, on causal priorities among them following the change program were examined.

Sophistication of technology was assessed using a questionnaire instrument constructed to measure the qualities of standard materials input, throughput mechanization, and output control for each work group. The questionnaire is described in Chapter III, above. This instrument utilized the structured judgments of work groups on nine scales by a small number of administrative people within each organization. A detailed analysis of the reliability and validity of this Technological Classification Form is presented in Taylor, 1970.

The planned social change program is an implicit independent variable in this study. This change program involved an attempted change in management values and behaviors in the direction of Likert's System IV (Likert, 1967).

Consultants from The University of Michigan introduced the planned change program. This effort began following the first questionnaire administration, and prior to the second one. It involved using the by-group results of the prior survey (or surveys, in the case of the refinery) as a self-help diagnostic tool, and specifically developed training programs as well. Since the consultants attempted uniform diffusion of inputs throughout the lower level ranks in each organization, this independent force is assumed constant.

Dependent variables. The ten variables, taken from the Survey Questionnaire, which were used in the analysis for the original study were used in the present replication as well. These variables fall into four classes: Four areas of supervisory leadership, four areas of work group or peer leadership, work group behavior, and satisfaction with the work group.

These measures formed the basis of the dependent variables—measures of the degree to which groups were originally participative and autonomous, and the extent to which they responded over time to the planned change program directed toward these ends.

Analysis design. The analysis took the form of controlling for high and low levels of technological sophistication and examining the dependent variables for differences in mean scores and, where possible, differences in cross-lagged, zero-order longitudinal correlations.

In order to keep the analysis simple, the data bases were combined such that groups with high sophistication of technological input, throughput, and output scores could be compared with groups with low input, throughput, and output scores. This total technology high-low design reduced the number of groups in each category, but in most cases still maintained reasonable N's for statistical purposes. The method of combining technology scores is that described as the "optimization approach" in Chapter IV above.

Results

Table 16 presents data relevant to the first hypothesis—that advanced technology by itself can influence autonomous group process. Table 16 presents mean scores, and differences between means for groups in the high-technology category versus groups in the low-technology category measured at time one (prior to planned change efforts). Tests of significance of difference between means were undertaken using the t-test statistic with two-tail evaluation.

Of the 39 differences between means presented in Table 16, 33 are differences where the high-technology category is higher in level of evaluation than the low-technology category (85 per cent of the total). Of these 33, 19 (49 per cent of the total) are large enough to generate t scores which are significant at the 10 per cent level of confidence or better. Of the six negative differences (where the low technology category is higher than the high technology condition) none attain significance at these levels.

With regard to specific dependent variables in Table 16, certain of them result in a similar pattern of strength across the four sites, and certain of them do not. For example, supervisory work facilitation and interaction facilitation, as well as peer goal emphasis and work facilitation, reveal consistent patterns of strong differences for high versus low technology, over the four sites. On the other hand, supervisory and peer support, as well as satisfaction with the group, show less consistency in strength of difference over the four sites. This is not unexpected since the effects of technology are assumed to be initially and primarily task related, and work facilitation and goal emphasis are task related leadership dimensions. The inconsistent results in support and satisfaction could well be a function of the incomplete diffusion of technological effects through confounding effects of continued use of older management patterns in the high-technology groups, and the substitution (or trade-off) of social for task leadership in low-technology groups.

In general, however, the differences between the mean scores for the high and low categories for each site at time one suggest that members of groups with high sophistication of production technology initially perceive higher levels of supervisory and peer leadership in their groups than did members of groups with low sophistication of technology. This is evidence that sophisticated technology, in and of itself, is associated with more autonomous and democratic group process.

Figures 6 through 11 provide summary information for a test of the second hypothesis. Figure 6 presents the average means for all four of the peer leadership variables for high- and low-technology categories in the refinery, compared with the combined four variables for the whole refinery (n = 2,200, N groups = 350), for the three measurement periods. It is clear from this figure that between times one and three, the high-technology groups start

Table 16

MEAN SCORE DIFFERENCES ON DEPENDENT VARIABLES (TIME ONE) BETWEEN HIGH AND LOW SOPHISTICATION OF TECHNOLOGY IN FOUR ORGANIZATIONS

Dependent Variables	Refinery (N = 44)			Glass Products (N = 18)			Plastics (N = 15)			Metal Products (N = 36)		
	High Tech.	Low Tech.	Diff.	High Tech.	Low Tech.	Diff.	High Tech.	Low Tech.	Diff.	High Tech.	Low Tech.	Diff.
Supervisory Support	3.92	4.14	-.22	4.21	3.35	.86**	4.29	3.64	.65*	3.44	3.23	.21
Supervisory Goal Emphasis	3.94	3.75	.19	4.15	3.05	1.10**	3.99	3.34	.65*	3.51	3.36	.15
Supervisory Work Facilitation	3.45	3.20	.25	3.68	2.65	1.03**	3.12	2.70	.42*	2.95	2.66	.29#
Supervisory Interaction Facilitation	3.64	3.23	.41*	3.75	2.72	1.03**	3.32	2.61	.71*	3.11	2.82	.29
Peer Support	3.83	4.03	-.20	3.78	3.72	.06	3.97	3.70	.27*	3.57	3.50	.07
Peer Goal Emphasis	3.55	3.44	.11	3.73	2.98	.75**	3.31	3.03	.28	3.29	3.09	.20#
Peer Work Facilitation	3.47	3.11	.36*	3.38	2.70	.68**	3.22	2.63	.59**	3.19	3.11	.08
Peer Interaction Facilitation	3.37	2.68	.69**	3.58	3.00	.58*	3.10	2.70	.40	2.96	2.99	-.03
Group Process	3.68	3.17	.51**	3.54	3.40	.14	-	-	-	3.15	3.37	-.22
Satisfaction with Work Group	4.38	4.22	.16	3.97	4.07	-.10	3.98	3.78	.20	3.91	4.04	-.13

**p < .01 two tail.
*p < .05 two tail.
#p < .10 two tail.

Figure 6

COMBINED PEER LEADERSHIP MEAN SCORE CHANGE OVER TIME FOR
HIGH- AND LOW-TECHNOLOGY GROUPS COMPARED WITH TOTAL COMPANY

Three Points in Time
(Refinery Data)

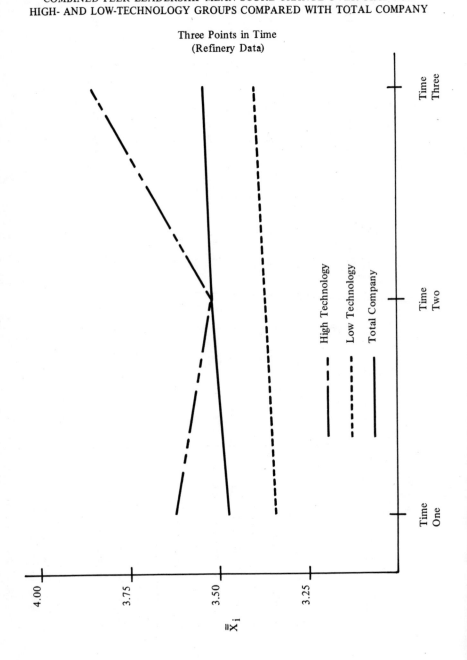

Figure 7

COMBINED PEER LEADERSHIP MEAN SCORE CHANGE OVER TIME FOR
HIGH- AND LOW-TECHNOLOGY GROUPS COMPARED WITH TOTAL COMPANY

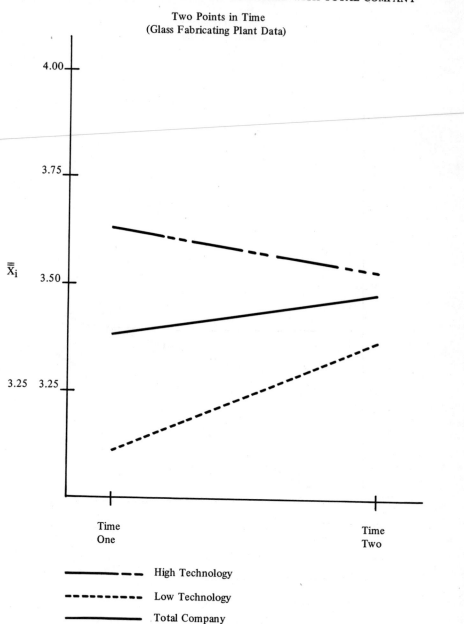

Two Points in Time
(Glass Fabricating Plant Data)

Figure 8

COMBINED PEER LEADERSHIP MEAN SCORE CHANGE OVER TIME FOR
HIGH- AND LOW-TECHNOLOGY GROUPS COMPARED WITH TOTAL COMPANY

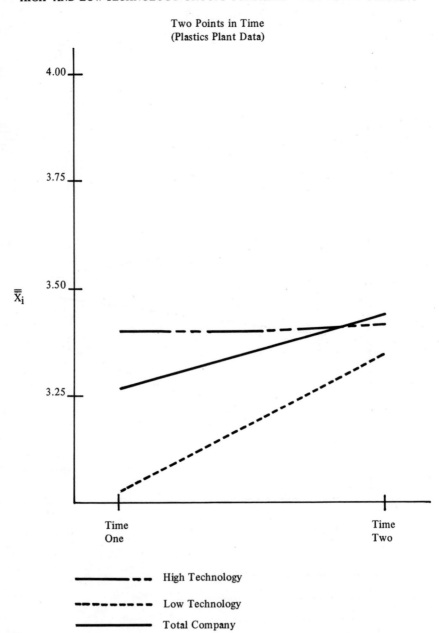

Two Points in Time
(Plastics Plant Data)

Figure 9

COMBINED SUPERVISORY LEADERSHIP MEAN SCORE CHANGE OVER TIME FOR
HIGH- AND LOW-TECHNOLOGY GROUPS COMPARED WITH TOTAL COMPANY

Three Points in Time
(Refinery Data)

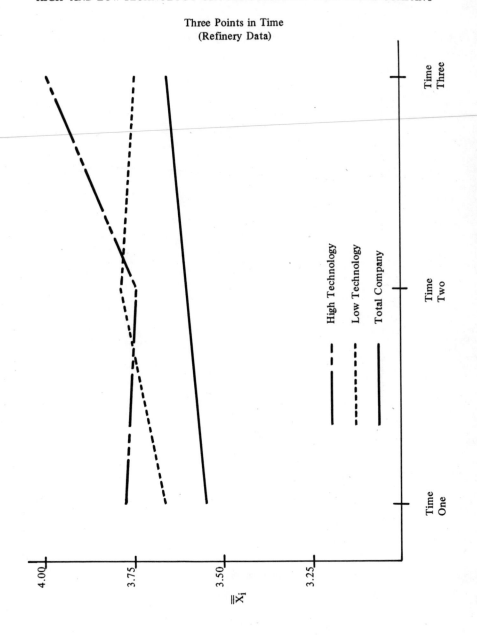

Figure 10

COMBINED SUPERVISORY LEADERSHIP MEAN SCORE CHANGE OVER TIME FOR
HIGH- AND LOW-TECHNOLOGY GROUPS COMPARED WITH TOTAL COMPANY

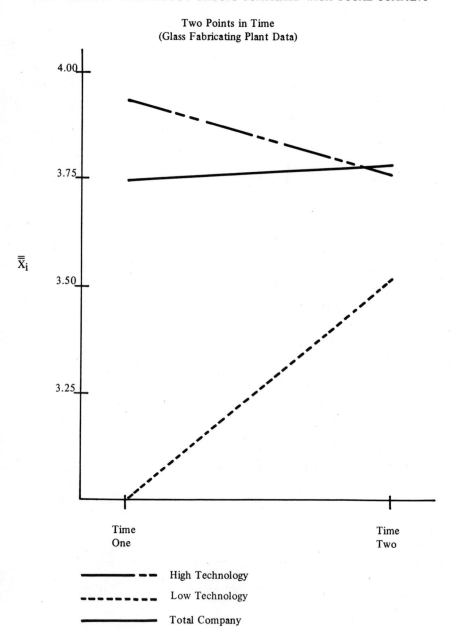

Two Points in Time
(Glass Fabricating Plant Data)

Figure 11

COMBINED SUPERVISORY LEADERSHIP MEAN SCORE CHANGE OVER TIME FOR
HIGH- AND LOW-TECHNOLOGY GROUPS COMPARED WITH TOTAL COMPANY

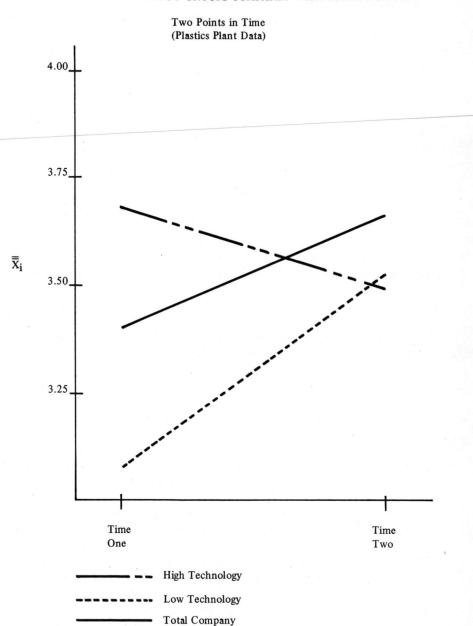

Two Points in Time
(Plastics Plant Data)

higher in peer leadership and increase faster than either the low-technology groups, or the refinery as a whole. The differences between t_1-t_3 are statistically significant for the high-technology category, and not significant for the low-technology category.

Figure 7 presents the average means for all four of the peer leadership variables for high- and low-technology categories in the glass fabricating plant compared with those variables for the whole glass plant (n = 332, N groups = 39), for the two measurement periods available. Figure 8 presents the same variables and groups using the plastics plant data compared to the total plant (n = 429, N groups = 71).

It is noted from Figures 7 and 8 that a remarkable resemblance exists between the two of them, and between them and the refinery data t_1-t_2 (Figure 6). In effect, what can be said is that between time one and time two, groups in the high-technology category decrease in evaluation of peer leadership, while groups in the low-technology category follow the trend of general increase manifest for the total company in each case. Thus, whatever happened in the refinery between t_1 and t_2 seems to be happening also in the other two plants. The phenomenon this pattern might reflect will be described below.

Figures 9-11 use the same data bases but present average means of the four supervisory leadership variables for the three sites involving subsequent measurement.

Once again, there is a statistically significant shift for the refinery, high-technology category between t_1-t_3, while that t_1-t_3 difference is not significant for the low-technology category (Figure 4). The time one-time two similarities among the three sites for peer leadership (Figures 6-8) are replicated for supervisory leadership (Figures 9-11). In all three cases, supervisory leadership decreases sharply for the high-technology condition t_1-t_2 and increases to a sizable degree for the low-technology condition. In all three sites, also, the total plant shows more gradual increase in supervisory leadership.

The following data help in examining the difference in causal effects in the high- and low-technology categories over time. Partial assessment of such effects were obtained using cross-lagged analysis of average zero-order correlations (Pearson r, or Spearman rho), while controlling for high and low technology.

Figures 12 and 13 present the dominant chains of causal priority among the dependent variables for the measurement periods available for the refinery and plastics plants respectively. The causal estimates shown in Figure 12, the refinery, were presented in Chapter IV, as Figure 5, and are duplicated here for ease of comparison. The causal estimates in Figure 8, the plastics plant, are presented for comparison with the refinery. Although the N of groups in the plastics plant analysis is quite small, these data, as a part of the replication effort, were felt useful and minimally satisfying statistical requirements. Although a similar analysis of causal estimates using the glass plant data was desirable, the N of groups in the low-technology category for that site was simply too small to be used.

In both Figures 12 and 13, an effort at simplification resulted in the removal of cases of reciprocal causality in favor of presenting only recursive, or more intransitive causal chains.

For the high-technology condition in the refinery (Figure 12), peer leadership (probably our best indicator of autonomous group process) time two is clearly the recipient of causal influence of the time one variables: group process, satisfaction, and, of course, itself. Supervisory leadership time two is influenced by peer leadership and group process time one. Nothing strongly influences satisfaction with the work group at time two. But by time three, consolidations in time two group process and peer leadership have led to realignments in satisfaction such that it is higher in groups which were more autonomous at time two. Thus, we may tentatively conclude that changes in group behavior, under the high-technology condition, tend to maintain themselves via changes toward consonant attitudes over time.

In the low-technology condition, originally *lower* group process tends to lead to *higher* satisfaction at time two and this, in turn, has a slight inverse causal influence on group process time three. It may be said that this situation reflects the lesser constraints on behavior for autonomous group process and little motivation for changing toward it. *Lower* group process time one produces *higher* satisfaction which, in turn, has a slight perseverative effect on low group process time three, where technological sophistication is low.

For the high technology in the plastics plant (Figure 13), a pattern is shown which is quite similar to that found in the refinery. Higher peer leadership t_1 leads to high supervisory leadership t_2. Peer leadership t_2 in the plastics plant, like the refinery, is influenced by satisfaction with the work group and itself t_1 (the group process variable shown for the refinery was not used in the plastics plant). A rather dramatic effect for the high-technology category in the plastics plant is the reverse influence of supervisory leadership t_1 on itself t_2.

The low-technology category in the plastics plant shows the stronger effects of supervision on subsequent peer leadership where behavior constraints of technology in the direction of autonomous group process are not manifest.

Discussion

We may conclude, with respect to the first hypothesis tested here, that technological sophistication in the four industrial sites initially measured prior to planned change input does have a measurable association with democratic and autonomous group process. Table 16 reveals quite consistently that high-

Figure 12

RESULTS OF CROSS-LAG CAUSAL ANALYSIS

Refinery Data

High Technology (N groups = 42)

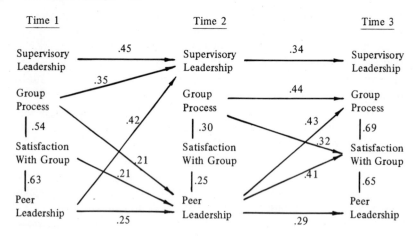

Low Technology (N groups = 26)

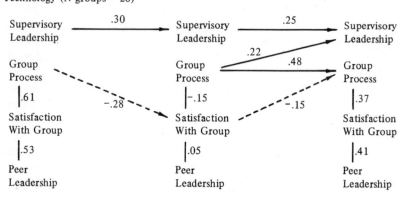

technology groups are initially higher in supervisory and peer leadership than low-technology groups before the planned change program got underway. These static relationships do not inexorably lead to statements of causality, but it seems likely that pre-change levels of autonomous group process were caused by, rather than caused, the advanced technology. This, of course, does not rule out a possible third variable leading to pre-change levels in both technology

Figure 13

RESULTS OF CROSS-LAG CAUSAL ANALYSIS

Plastics Plant Data

High Technology (N groups = 8)

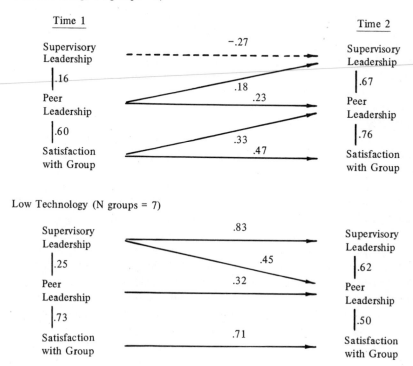

and leadership. It is unlikely, however, that something as direct as demographic characteristics are operating on both major variables. A description of the rather detailed examination of the effects of age, education, tenure, and place of respondent upbringing (urban-rural) on the evaluation of technology, and on the dependent variables in the refinery (Chapter IV) revealed minimal to non-existent influence of these characteristics as confounding variables.

In regard to the second hypothesis, that sophisticated technology facilitates social change in the direction of autonomous group process, the results obtained are remarkably consistent. As Figures 6 through 11 revealed, groups in the high-technology category began at time one with higher mean scores in both peer and supervisory leadership in all three sites where questionnaire measurement was obtained subsequent to the onset of a planned change program. Between time one and time two in these three sites, the level of peer leadership in the high-technology condition decreases slightly or stays the same, while the level of supervisory leadership decreases faster. Low-technology groups in all three sites increase in supervisory and peer leadership during the same period. The refinery data (the only site with three measurement periods presented here) between time two and time three, however, show a marked increase in leadership which amounts, overall, to a greater increase t_1-t_3 for high-technology groups than for the low-technology groups. The possibility of "gain effect," rather than technology itself, affecting this increase was tested and dismissed in Chapter IV (c.f., Table 15).

What might account for the decrease in high-technology leadership in all three sites between time one and time two is a peculiar interaction among the strategy of the planned change program introduced in all three sites, the time-frame between measurements, and the behavioral constraints of advanced technology.

It has been noted above that, in all sites, the planned change program involved the feedback of work group data to individual foremen and supervisors. These data were to be used by these people to suggest ways that they might improve in the direction of participative management. Change agents worked closely with these foremen and supervisors in discussing the survey results with subordinates, and in determining what training might be undertaken to improve participative activities on the part of subordinates. This strategy perforce implies that the supervisor is the keystone—once he has decided to attempt change, he will carry that change to his subordinate work group. The advanced technology, on the other hand, places constraints directly on the work group to behave in a more autonomous and participative way, and acts only indirectly on the supervisor or foreman in forcing him to distance himself from the activities of his subordinate group.

What the t_1-t_2 curves in Figures 6-11 may be revealing is the effect of what can be considered a less than optimal input by the foreman or supervisor in his desire to exercise more participative management. This input on the part of the supervisor could well be seen as disruptive to a high-technology work group where a relatively high level of peer leadership is initially manifest. If, as we may speculate, the supervisor in a high-technology group is somewhat distant to his group prior to the planned change effort (and is seen to supervise well), his introduction of new ways of managing and his increased visibility may actually disrupt the relatively comfortable fit between behavior and technology developed by the work group. As Figures 6-8 show, high-technology groups either maintain initial levels of peer leadership or decrease a nonsignificant amount between t_1 and t_2. As Figures 12 and 13 clearly show, supervisory leadership in high-technology groups at time one does not cause peer leadership time two, contrary to what we might expect. In fact, in both cases peer leadership time one has causal influence on supervisory leadership time two where technology is sophisticated. These data suggest that the supervisor of a high-technology group, at the onset of the planned change program, attempts to influence his work group, but in fact is influenced by them. In the low-technology groups, on the other hand, the implicit change strategy seems to fit better with immediate increases in supervisory and peer leadership, and possibly t_1-t_2 unilateral influence of supervisor on peer leadership.

Since all three organizations reported here were measured over a 12-month period, a question of time-frame arises—namely, would it not be more reasonable to expect that similar changes in all three organizations would take place in the same period? In answer to this, it must be noted that, in all cases, the supervisors knew when the next questionnaire survey would be undertaken and what sort of deadlines they had for attempting change. In this light, the tighter time-frame of six months between the three measurements in the refinery might have collapsed or telescoped the change activities and concerns of group supervisors in the refinery and provided the opportunity for members of high-technology groups to help the supervisor understand what was required of him for more participative management in such a setting.

Examination of additional data in a search of an explanation of the t_1-t_2 changes for the high-technology groups is suggestive, but by no means definitive. If we can assume, as we have above, that immediately following onset of the planned change efforts, the supervisor of a high-technology group is seen by his subordinates to be more disruptive or meddling *vis-a-vis* group process, while low-technology supervisors are seen as helpful in this period, then we can speculate that supervisory behavior which is seen as desirable by subordinates at this time will differ between the two groups as well. Specifically, we would expect that ideal supervisory scores (how the respondent "would like" the supervisor to behave) would initially be closer to actual scores for the high-technology groups, and would remain more constant over time for the high-technology groups (indicating subordinate desire for the supervisor to continue what he had been doing initially).

Examination of ideal supervisory scores t_1-t_2 for high- and low-technology categories tends to support this. Between time one and time two in all three sites, ideal supervisory scores are initially closer to obtained scores for the high-technology groups and remain fairly stable over the period. In two of the three sites, the ideal scores for the low-technology category increase and at a faster rate than for the high-technology groups. Low-technology groups see considerable initial disparity between actual and ideal supervisory leadership. As low-technology groups see increase in supervisory behavior t_1-t_2, they also desire higher levels; high-technology groups, on the other hand, have a lower initial discrepancy between actual and ideal supervisory leadership and tend to maintain the level of desired supervisory behavior over time, hence creating greater disparity between desired and obtained leadership at time two. The effect of disparity between actual and ideal leadership on subordinate satisfaction with supervisor is fairly clear. For the high-technology condition, satisfaction with the supervisor tends to diminish between t_1-t_2, while it increases for the low-technology groups during that period. With regard to the refinery data, ideal supervisory leadership t_2-t_3 increases at the same rate as does actual leadership during that period, as does satisfaction with the supervisor. This last finding lends support to the notion that the supervisor learned how better to adapt to the high-technology condition following the second feedback activity.

Conclusions and Implications

Replication of the original study design in industrial organizations with smaller number of groups and fewer measurement waves tends to confirm the original conclusion that advanced technology, in and of itself, is associated with more autonomous and participative work group process. In three industrial cases (glass, metals fabrication, and plastics production), a pattern of greater peer leadership is manifest for groups with more sophisticated technology. This pattern parallels the one originally found and reported in Chapter IV for the petroleum refinery.

Another result obtained in the original refinery data, that advanced technology facilitates planned social change efforts directed toward more participative management, was not directly supported. Remarkable consistency in the pattern of social system change was found, however, between the original refinery data, t_1-t_2, and the data in two of the replication sites. Originally, in the refinery, it had been found that over three waves (t_1-t_3) in 12 months the high-technology category increased faster in peer and supervisory leadership than either the low-technology category or the refinery as a whole. Looking first at change between waves one and two (t_1-t_2), the high-technology condition decreased in leadership evaluation in the refinery, however. This t_1-t_2 pattern was replicated nearly exactly for the glass fabricating plant and the plastics producer, which were only measured twice, not three times, in a 12-month period.

In discussing this marked similarity among the t_1-t_2 results for the three plants, an explanation was advanced which asserts that this initial decrease in the high-technology condition might be caused by the particular change strategy employed in these plants. This strategy employs the group supervisor as the lever for change, and requires that he increase his activity level *vis-a-vis* his subordinate work group. Where the technology is advanced, it appears from the time one data that the supervisor is initially distant and the group relatively autonomous. In following the change program, and attempting to increase participative activities in his team, the supervisor of a technologically sophisticated group may be seen by his subordinates to be doing more interfering than helping. The changes noted by time three measurement (especially

t_2-t_3) in the refinery suggest that this interference may be transitory as the supervisor recognizes what is necessary in the eyes of his subordinate group and modifies his behavior appropriately. The problem of time-frame of measurement and possible effects on speed of change in the refinery as compared to the other two plants was discussed. It is concluded that the shorter dead-lines of six months between measurement and the opportunity for supervisors in the refinery situation to receive this additional feedback from subordinates might have speeded up the change process in high-technology groups. In any event, it is clear that between time one and time two in all sites measured, the planned change strategy used had apparently disruptive (or perhaps unfreezing) effects on subordinate perceptions of supervisory leadership and, to a lesser degree, on peer leadership. Additional research and data collection are required to study this possible effect and the longer-range facilitation noted in the refinery results.

VII

SUMMARY AND CONCLUSIONS

The purpose of the present research has been to test several specific notions advanced by socio-technical theorists regarding the ability of technical systems to facilitate planned social change efforts. Also included in the present study was an attempt to establish, using concepts popular in attitude change theory, that technology not only facilitates social change, but provides for more permanent change as well.

The reasons behind the study are two-fold. *First*, the socio-technical theory of industrial organization is an interesting and plausible theory which, in conceptual strength, is much beyond the main body of literature dealing with organizational response to technology which flows currently from a seemingly endless *Zeitgeist*. Although the socio-technical theory grew out of empirical data, those data are relatively limited in terms of the methods used to collect them. Unintentional corraborative data are available, but are not directly comparable. Therefore, an independent test of some predictions of socio-technical theory using different methods of study seems useful and interesting. *Second*, organizational change, its strategy and tactics, is a topic of considerable and lasting importance in our changing society. To the degree that the present study provides specific and convincing knowledge regarding one parameter of organizational change, namely technology, it will contribute to something more than the purely academic.

A rather extensive literature review was undertaken (Chapter I) to assess the degree to which studies of organizational response to production technology confirmed or supported the rather specific conclusion of the Tavistock socio-technical theorists that more sophisticated technology provides a force toward a democratic group style of leadership in organizational behavior. In the main, the other studies confirmed this conclusion, although the individual studies varied in the extent to which this result obtained or the degree to which absence of results could be explained as artifactual. It was concluded that the socio-technical position was strengthened in light of the other findings. It was also concluded that since the Tavistock evidence did not include white-collar industry, and because the additional literature did not show effects as strong for white-collar industry as those for blue-collar, control for this factor was needed. A brief review of organizational change literature was also undertaken in Chapter I, where an explanation of technological effects was discussed—an explanation based on specific and directional situational constraints by technology on human behavior. It was concluded that technology might not only facilitate social change, but might provide for more permanent change effects as well. This prediction was based on the discussion of situational forces or structural changes and their effect on attitude change via role changes in opposition to prevalent attitudes held by the role occupant. Finally, it was also noted in Chapter I that little had been attempted in obtaining measures of what sophisticated technology was. Each research study cited in the literature review used a different, if not always specified, method of technological assessment. This, it was felt, might account for much in the inconsistent nature of the research findings. It seemed necessary, given this lack, to attempt to develop a replicable measure of technological sophistication, one applicable to many different kinds of organizations, in order to obtain more valid research results in the present study.

The thesis that sophisticated technology facilitates planned social change and results in more permanent effects was tested using data from studies undertaken in several industrial and commercial organizations. The main study (Chapters III-V) involved the use of data taken from studies of a medium-sized insurance company, and of a large petroleum refinery. The responses of 2,800 persons in nearly 300 work groups were used in the present main study. The replication study (Chapter VI) used data collected in three other industrial organizations. These included a glass products factory, a continuous-process plastics producer, and a metal products manufacturer. Together these sites represented the responses of almost 1,400 persons in over 200 work groups. The samples used in the original study were of large size and they both included three data collection periods. The replication sites on the other hand were smaller and were sampled no more than twice. These characteristics made the replication study data both too small for some tests undertaken in the main study, and too limited in sample periods for other tests.

Beside the general hypothesis that technology facilitates planned change efforts, several specific predictions were proposed. One of these was a prediction regarding technological effect on work group behavior before change attempts were undertaken. The remaining hypotheses were phrased as specific predictions from the general hypothesis and dealt with causal relationships among the questionnaire variables and between the questionnaire variables and the technology variable. These predictions fell into three categories: influence of early supervisory leadership and technological sophistication in planned change, technological sophistication and self-maintenance of peer leadership in planned change, and the reinforcing effects of changed attitudes on subsequent behaviors in peer leadership and group process under sophisticated technology during planned change.

Tests of the general hypothesis revealed confirming results. In the refinery, the high-technology group over time improved more in participative peer and supervisory leadership, and group process than did the low-technology group. In the insurance company, where a general condition of shift in mean questionnaire response over time in a direction away from participative management was noted, the high-technology group means shifted at a slower rate than did the means in the low-technology group; the high group seemed more resistant to external change forces in the direction of less participation than did the low group.

In terms of the first specific hypothesis, it was found that the results held for the refinery and all three of the replication sites, but not for the insurance company. Except in the insurance company, mean scores for peer leadership and group process reflected greater democratization for high-technology groups than for the low-technology groups before the change efforts were undertaken. In the insurance company, reverse results were obtained which, it was suggested, were probably a function of education.

Overall, results tended to provide some support for the remaining set of predictions. The expected results of early supervisory leadership and technological sophistication causing subsequent peer leadership and group process were obtained in part. It was also noted, however, that in the absence of additional measurement between times one and two, a better model might be that peer leadership might be a predictor for supervisory leadership in the high-technology condition.

The predictions requiring three measurement periods were generally confirmed in the refinery, and remained untestable in the insurance company. Peer leadership had strengthened enough by time two in the high-technology condition in the refinery that it could maintain itself and influence subsequent group process.

Another set of predictions were also confirmed in the refinery sample, and were basically untested in the other sites. These results supported the notion that situational constraint in sophisticated technology plays a role in establishing early permanence of favorable change via the reinforcing effects of attitudes on behaviors.

The overall results obtained in refinery and insurance company suggested that technology acts both directly on behavior and indirectly on system strength as a conditioning variable on the relationship between planned change efforts and effects of planned social change. An unexpected finding was that technology has an effect on the interrelationships among the intragroup social subsystems measured in the present study, which seems independent of similar effects found using education. That this interconnectedness, or system strength, is related to degree of change was also found, and is not an unlikely event.

Finally, results from the replication study confirmed a pattern originally found in the refinery that levels of leadership, both supervisory and peer group, decreased in high technology groups between time one and time two. The compelling consistency of these results demanded a post hoc explanation which was put forward in Chapter VI as the effect of an interaction among the strategy of the planned change program introduced in all sites, the time-frame between measurements, and the behavioral constraints of advanced technology. The existence of this interaction brings into question the universal application of a "top down" change strategy which relies on the group supervisor acting as the sole leverage for change in his subordinates. It also highlights the possible importance of setting deadlines for stages in the organizational change program which are clear and understood to all members of that organization.

Conclusions

The results of the research described above offer the following specific conclusions:

1) A measure of sophistication of production technology was developed which had reasonably high inter-rater reliability and factorial and convergent validity. To the degree that the results of this measurement instrument fit the theoretical predictions, it can also be said to have reasonably good construct validity. This measure appears to be sensitive to differences in white-collar as well as blue-collar industries.

2) The measure of technological sophistication distinguished between groups with different pre-change levels of subordinate perceptions of supervisory and work group behaviors.

3) Technological sophistication does facilitate or enhance change forces in the direction of participative management or autonomous group functioning, *and* also seems to attenuate change forces in the direction of less participative, or more authoritarian management.

4) Technological sophistication seems to operate as a conditioning variable in social change efforts both directly through situational constraint on worker behavior, and indirectly through affecting interconnectedness of social subsystems.

5) Technological sophistication acts to increase permanence of change efforts by providing a situation where changes in attitudes are strong subsequent effects of changed behaviors. These changed attitudes appear to be reinforcing factors in the continuance of the changed behaviors.

The present study, it is felt, provides substantial contributions and research strength to the socio-technical literature:

1) It integrates and tests in one study many implicit hypotheses previously scattered throughout the literature.
2) It provides replication of many findings across a number of similar organizations.
3) It provides additional longitudinal and quantitative measurement to a field containing its share of static correlational studies and rather subjective case studies.
4) Finally, it provides promising bases for application and further research in the study of planned change.

APPENDIX A

PART I – SOPHISTICATION OF INPUT

1) Standardization of material (i.e., the objects transformed by the group)

: 5 :	4 :	3 :	2 :	1 :
The materials or the objects transformed by employees are treated in a totally standardized or routine fashion.	Objects transformed are treated as containing mostly standardized elements.	The objects transformed are considered partly standardized or routine. That is, the average or normal object can be considered partly standardized and partly unique.	Objects transformed are treated as being composed of mostly unique elements.	Objects transformed are treated as virtually unique entities, one from another.

2) Predictability (of objects transformed) in those characteristics important to production operations

: 5 :	4 :	3 :	2 :	1 :
When working with the normal case in this group, employees *nearly always* experience certainty and predictability in the objects they are transforming or manipulating.	Average employees *usually* experience certainty in the input or objects they are transforming in the average case.	Employees consider that raw materials in the normal case are sometimes predictable and sometimes not.	Employees in this group *usually* consider the input or material transformed in the average case as uncertain and unpredictable.	The objects transformed in the average case are *always* considered unpredictable by the employees.

3) Knowledge of raw material enabling analytical approach to problems.

: 5 :	4 :	3 :	2 :	1 :
So much is known about the objects transformed that employees in the group can approach nearly all non-routine problems in a logical, analytical way.	Enough is known about the raw material, or objects transformed so that the average employee can resolve many exceptions in a systematic, logical way.	Something is known about the objects transformed which enables employees in this group to resolve some non-routine problems in a systematic analytical way.	A little is known about the raw material so that employees can resolve a few exceptional cases in an analytical, systematic way.	So little is known about the objects transformed that employees in this group cannot be expected to approach any non-routine cases in an analytical way.

127

PART II – SOPHISTICATION OF MACHINES – THROUGHPUT

4) The *proportion* of routine operations which are not handled by machines.

1	2	3	4	5	0
The jobs in this group contain a great proportion of routine operations which although it is presumed may be handled by machines, are not.	These jobs have a sizable proportion of such routine operations which could be taken care of by machines and are not.	The jobs in this group have a moderate proportion of routine operations which have not been mechanized.	This group performs jobs which contain a low proportion of routine operations which have not been mechanized.	These jobs include virtually no routine operations which have not been mechanized.	These jobs originally had a low proportion, or no routine operations in a non-mechanized state, so they've had none to mechanize.

5) Predominant type of mechanical operation—regardless of how much or little machine operation is involved in the job.

1	2	3	4	5
Mechanical operations performed in jobs in the group are primarily via human effort: manual tools, nonautomatic, nonpower-assisted.	Most machines in the group are power-driven but not automatic (i.e., very few routine operations are controlled automatically).	Power-driven tools or machines with automatic control of *some* routine operations account for the typical mechanical operation.	Power-driven machines with automatic control of *all* routine operations is the typical mechanical vehicle in this group.	The predominant mechanical device in this group is power-driven machines with automatic and programmatic control of most unexpected or nonroutine, as well as routine operations.

6) The extent the machine is dependent upon the operator—when that machine is being used. (The machine as servitor)

1	2	3	4	5	0
The employees cannot leave their machines or work place at all during any portion of the task without stopping work.	The employee can leave his machine while it is running for only two or three minutes at a time.	The employees operate a machine with continuous feed from some distance, but cannot leave the control equipment for more than a few minutes while the machine is running.	The employee operates a machine which can perform a production operation itself for longer than a few minutes and will shut down automatically in the event of unprogrammed disruption.	The employee operates a completely automatic machine with continuous feed which will shut down by itself in the event of trouble.	The employees utilize machines to a very limited or no extent.

PART III – SOPHISTICATION OF FEEDBACK – OUTPUT

7) The supervisor as a source of evaluative feedback.

:	1	:	2	:	3	:	4	:	5	:
	The supervisor can provide close or *nearly continuous* watch, help, and evaluation of the quality of performance of the average employee in this group.		The supervisor can provide some technical help and *daily* evaluation of the quality of performance.		Occasional evaluation of performance quality (*less than daily*) and technical help in resolution of non-routine problems can be provided by the supervisor.		The supervisor can provide occasional supplemental evaluation of employee performance quality (*less than weekly*) and provides some technical help.		Except for annual, or other periodic appraisals, the supervisor cannot provide evaluation of employee performance, or technical help.	

8) The supervisor as an initiator of evaluative feedback.

:	1	:	2	:	3	:
	The supervisor usually *initiates* what feedback he provides to the average employee in this group.		The supervisor *initiates* evaluative feedback about half the time, and the other half, it is *requested* by the employees in this group.		The supervisor usually responds with the evaluative feedback and technical help he provides on *request* of the average group member.	

9) Time lag in feedback.

:	5	:	4	:	3	:	2	:	1	:
	In this group, the average time lag between performance and evaluative feedback is usually *longer than 16-18 hours* after completion of the article.		The time lag between performance and evaluative feedback falls usually *within the same day* the article is finished.		Feedback following performance of the average task in this group is usually *within an hour* of completion.		The time lag of feedback following performance for the average task in this group is usually *within a few moments* of completion of each article.		The time lag between performance and evaluative feedback for the average task in this group is *nearly instantaneous.*	

10) The type of primary source(s) of evaluative feedback.

:	2	:	1	:
	The primary source(s) of feedback for the average job in this group is (are) *human* – the supervisor, or work group members, quality control engineers, or other persons in associated groups.		The primary source(s) of feedback in this group is (are) *non-human* – e.g., dials, guages, measuring devices, reports, or printed statements, audible signals, etc.	

PART IV – CONTROL ITEMS – EFFECTIVENESS

*If the technology or work system in the group is new or fairly new
(within the past four or five years) answer the following question.*

11) What sort of expectation did the organization have for the performance of the technology or work system in this group?

: 1 :	2 :	3 :	4 :	5 :	9 :
It was not expected to do any more than the preceding system.	It was expected to do a *marginal* amount better (in quality, quantity, or cost)—perhaps 10% better.	It was expected to do *somewhat* more than its predecesor—about 50% better.	It was expected to do *a great deal* better in quantity, quality, or cost—perhaps twice as good.	It was expected to be a *major break-through* in this sort of work—possibly five times improvement.	This question is not applicable to this group.

*If the technology or work system in the group is new or fairly new
(within the past four or five years) answer the following question.*

12) To what extent is the new work system used in this group effective? (To be effective the system must do what it was supposed to do—e.g., is it providing more and/or better output with the same number of people, or the same output with fewer human interventions?)

: 1 :	2 :	3 :	4 :	5 :	9 :
It is *very ineffective*—not operating nearly as well as expected.	It is *somewhat ineffective*—not operating quite as well as expected.	It is *effective*—operating as expected.	It is *very effective*—operating somewhat better than expected.	It is *extremely effective*—operating far better than expected.	This question is not applicable to this group.

*If the technology or work system in the group is new or fairly new
(within the past four or five years) answer the following question.*

13) How long did it take to complete the start-up period for the installation of this system?
That is, what was the period before the system operated as well as could be expected?

: 0 :	1 :	2 :	3 :	4 :	5 :	9 :
It hasn't yet, but it's too soon to tell if things will improve.	It never did and probably never will operate well.	It took *8-24 months* before it operated well.	It took *6-8 months* before it operated well.	It took *3-5 months* before it operated well.	It took *1-2 months* before it operated well.	This question is not applicable to this group.

INSTRUCTIONS FOR COMPLETING THE

"TECHNOLOGICAL CLASSIFICATION FORM"

We at The University of Michigan's Institute for Social Research are interested in the development of a scheme by which the technology of direct productive functions can be assessed for many different types of industries—white-collar as well as blue-collar. These rating forms you have been asked to complete are an initial attempt to achieve this. They represent an accumulation of knowledge obtained over the last decade in the study of technological systems and in their present form are synthesis of what we believe to be the crucial elements.

This rating scheme bears a strong methodological resemblance to the more familiar job description rating procedures used for setting job-specific wage rates in organizations. Experience in job description rating has established that the reliability of the measures is enhanced by providing the raters with commonly-known job groups. We are, therefore, including examples to be reviewed before undertaking the rating of the groups or functions listed on the rating sheets themselves.

Immediately following, you will find brief descriptions of several imaginary work groups (pp. 2-3). These are followed by evaluations of these groups on the same scales you will be asked to use (pp. 4-7). Read through these materials as carefully as you need to in order to become familiar with the sort of issues and questions you are likely to face.

After familiarizing yourself with these examples, please read the instructions entitled "Your Rating Task" which follows them.

DESCRIPTIONS OF TYPICAL WORK GROUPS

1. Secretarial Services—The ABC Insurance Company (abbreviated "SS" in evaluation to follow).

This group is composed of four transcription typists, and six copy typists. Employees in the former category transcribe correspondence reports and memos from dictating equipment, while those in the latter category type other materials, primarily policy endorsements from rough copy. The supervisor distributes the work to the typists according to the type of work and current load. The supervisor also evaluates the work and provides the brief training and technical help her girls require. Finally, the supervisor acts in the capacity of communication link between this group and the rest of the organization.

2. Seamless Tube Mill (abbreviated "STM" in evaluation to follow).

Three shift crews handle the 24-hour operation of this automated tube mill in the Ajax Steel Company: They are composed of eight men plus a foreman for each shift for a total mill team of 24 men. This mill, in five operations, converts billets of steel into a seamless tube of any given diameter. This new machine has been in operation for four years and was running smoothly after a 30-month period. The role of the foreman has evolved such that the workers are left alone to run the equipment pretty much themselves. The foreman acts in the capacity of obtaining materials, coordinating maintenance services, administering wages, and in general acting as the link to the organization on all but technical matters. Engineers and higher management still on occasion contact the workers directly regarding technical matters. The men on a crew must work together to keep the machine running although, since they are not located in one place, they communicate over a public address system. Of the eight jobs on each shift only two still require careful constant attention and repetitive activity during operation. These are located at the machine itself. The other six are primarily performing tasks at dials and controls which are variously located along the work flow some few feet from the equipment.

3. Cafeteria Kitchen Staff—Gas and Electric Company, Main Office (abbreviated "CK" in evaluation to follow).

This cafeteria serves two meals a day to 400 people as well as being open for coffee breaks. The kitchen is operated by five employees in addition to the supervisor-nutritional advisor who, when time allows, also works on food preparation. The staff includes a hot cook, a fry cook, a combination dishwasher and salad man, and two persons who serve from the steam tables and otherwise help in preparation. The kitchen has much modern equipment such as electric mixers, and potato peelers, an automatic dishwashing line, and thermostatically controlled deep fryer, fry grill, and steam table. Much prepared food, commercially partial-cooked food, and frozen food is used in the preparation of meals. The supervisor plans all meals, sets prices, does the buying and closely supervises all operations. The supervisor acts as the communication link between the kitchen and dining room staff (busboys, cashiers and janitors) although her linkage with the rest of the organization is through her immediate supervisor, the head of personnel services.

4. Safety Engineering Group. (abbreviated "SE" in evaluation to follow).

This is the safety engineering group for the Alpha Division of the Delta Manufacturing Company, a large producer of durable consumer goods. The group includes five industrial engineers, a statistical clerk and one secretary, in addition to the supervisor. Of the engineers, two are statistically oriented and handle much of the compilation of data and its analysis. The other three are more concerned with problems on the shop floor, such as the specific activation of safety modifications and techniques, the collection of information, routine and special, and its dissemination in the plant. Because of the size of the Alpha operation, much of the input data was computerized several years ago and the engineers use the computer to additionally analyze the data. Small, special treatments of the data are usually produced by hand by the statistical clerk under the direction of an engineer. The supervisor coordinates the efforts of his people and occasionally supervises their work. He receives most initial contacts from the rest of the organization, and represents the efforts of the group as well.

YOUR RATING TASK

The rating sheets that follow are arranged so that one question or scale is rated for all groups at a time. The scale is placed at the top of the page and the groups are arrayed below it. Please evaluate all of the listed groups or functions for that scale before moving on to the next question or scale.

If you do not know a work group you may skip it on each page, but please do so *only* if you are quite unfamiliar with it. If you have answered some of the questions for a particular group, please try to check a response to all of the questions for that group. It is expected that not all judges will be able to evaluate all groups, but it is presumed that you have been provided with evaluation forms dealing with groups that you should know for these purposes. If a group contains many jobs which are different with regard to any particular scale or question, try either to draw an average for the group, or to choose the most common or usual activities which are relevant to the scale.

It is important that you think about the following groups in terms of the *system* rather than the *persons* in all cases. For example, you will find questions about the role of the supervisor in providing evaluative feedback. Think about the *system's constraints* on the supervisor as a replaceable unit, in providing such feedback—*NOT* about the personal characteristics of the particular supervisor who now fills the role.

(For the administration, the following examples actually precede the section "Your Rating Task" above.)

1) Standardization of material (i.e., the objects transformed by the group).

: STM [1]	: SS / CK	: SE [2]	:	:	:
The materials or the objects transformed by employees are treated in a totally standardized or routine fashion.	Objects transformed are treated as containing mostly standardized elements.	The objects transformed are considered partly standardized or routine. That is the average or normal object can be considered partly standardized and partly unique.	Objects transformed are treated as being composed of mostly unique elements.	Objects transformed are treated as virtually unique entities, one from another.	

2) Predictability (of objects transformed) in those characteristics important to production operations.

: STM	: CK / SS	: SE	:	:
When working with the normal case in this group, employees *nearly always* experience certainty and predictability in the objects they are transforming or manipulating.	Average employees *usually* experience certainty in the input or objects they are transforming in the average case.	Employees consider that raw materials in the normal case are sometimes predictable and sometimes not.	Employees in this group *usually* consider the input or material transformed in the average case as uncertain and unpredictable.	The objects transformed in the average case are *always* considered unpredictable by the employees.

3) Knowledge of raw material enabling analytical approach to problems.

: STM / SS	: SE	: CK	:	:	:
So much is known about the objects transformed that employees in the group can approach nearly all non-routine problems in a logical, analytical way.	Enough is known about the raw material, or objects transformed so that the average employee can resolve many exceptions in a systematic, logical way.	Something is known about the objects transformed which enables employees in this group to resolve some non-routine problems in a systematic analytical way.	A little is known about the raw material so that employees can resolve a few exceptional cases in an analytical, systematic way.	So little is known about the objects transformed that employees in this group cannot be expected to approach any nonroutine cases in an analytical way.	

1. This evaluation of the Tube Mill is extreme since the operators themselves may see small differences in even the most standardized steel, but we *assume* the stock is *very* stable.

2. These safety engineers use standardized accident records and tabulation, but they also use specialized data they themselves collect.

4) The *proportion* of routine operations which are not handled by machines.

:	:	:	CK / SE	:	STM / SS	:	:	:
The jobs in this group contain a great proportion of routine operations which al-though, it is presumed may be handled by machines, are not.	These jobs have a sizable propor-tion of such routine opera-tions which could be taken care of by machines and are not.	The jobs in this group have a moderate propor-tion of routine operations which have not been mechanized.	This group performs jobs which contain a low proportion of routine opera-tions which have not been mechanized.	These jobs include virtually no routine operations which have not been mechanized.			These jobs origi-nally had a low proportion, or no routine operations in a non-mechanized state, so they've had none to mechanize.

5) Predominant type of mechanical operation—regardless of how much or little machine operation is involved in the job.

:	:	CK / SS	:	SE [3]	:	STM	:	:	:
Mechanical operations per-formed in jobs in the group are primarily via human effort: manual tools; non-automatic, nonpower-assisted	Most machines in the group are power-driven but not auto-matic (i.e., very few routine operations are controlled automatically).	Power-driven tools or machines with automatic control of *some* routine operations account for the typical mechani-cal operation.	Power-driven machines with automatic control or most routine operations are the typical mechani-cal vehicle in this group.	The pre-dominant mechanical device in this group is power-driven machines with automatic and program-matic control of most un-expected or nonroutine, as well as routine operations.				

6) The extent the machine is dependent upon the operator—when that machine is being used. (The machine as servitor)

:	SS	:	CK	:	STM	:	:	:	SE [4]	:
The employees cannot leave their machines or work place at all during any portion of the task with-out stopping work.	The employee can leave his machine while it is running for only two or three minutes at a time.	The employees operate a machine with continuous feed from some distance, but cannot leave the control equip-ment for more than a few minutes while the machine is running.	The employee operates a machine which can perform a production operation itself for longer than a few minutes and will shut down automatically in the event of unprogrammed disruption.	The employee operates a com-pletely auto-matic machine with continuous feed which will shut down by itself in the event of trouble.				The employees utilize machines to *a very limited* or no *extent*.

3. This is a combination or "averaging" of computers, calculators, etc., in safety engineering.

4. Since safety engineers use of machines or tools is primarily indirect, this evaluation is made.

7) The supervisor as a source of evaluative feedback.

: CK / SS :	SE [5]	: SE :	: STM :	:
The supervisor can provide close or *nearly continuous* watch, help, and evaluation of the quality of performance of the average employee in this group.	The supervisor can provide some technical help and *daily* evaluation of the quality of performance.	Occasional evaluation of performance quality (*less than daily*) and technical help in resolution of non-routine problems can be provided by the supervisor.	The supervisor can provide occasional supplemental evaluation of employee performance quality (*less than weekly*) and provides some technical help.	Except for annual, or other periodic appraisals, the supervisor cannot provide evaluation of employee performance, or technical help.

8) The supervisor as an initiator of evaluative feedback.

: SS :	CK :	STM / SE :
The supervisor usually *initiates* what feedback he provides to the average employee in this group.	The supervisor *initiates* evaluative feedback about half the time, and the other half it is *requested* by the employees in this group.	The supervisor usually responds with the evaluative feedback and technical help he provides on *request* of the average group member.

9) Time lag in feedback.

: SE :	STM / SS :		: CK :	
In this group, the average time lag between performance and evaluative feedback is usually *longer than 16-18 hours* after completion of the article.	The time lag between performance and evaluative feedback falls usually *within the same day* the article is finished.	Feedback following performance of the average task in this group is usually *within an hour* of completion.	The time lag of feedback following performance for the average task in this group is usually *within a few moments* of completion of each article.	The time lag between performance and evaluative feedback for the average task in this group is *nearly instantaneous.*

10) The type of primary source(s) of evaluative feedback.

: CK / SS :	STM / SE :
The primary source(s) of feedback for the average job in this group is (are) *human*—the supervisor, or work group members, quality control engineers, or other persons in associated groups.	The primary source(s) of feedback in this group is (are) *non-human*—e.g., dials, guages, measuring devices, reports, or printed statements, audible signals, etc.

5. It is assumed that judges here are thinking of the *technology's constraints* on the supervisor's ability to closely observe and help.

"SUPPLEMENTARY QUESTIONS FOR RECENT INNOVATIONS" [6]

If the technology or work system in the group is new or fairly new
(within the past four or five years) answer the following question.

11) What sort of expectation did the organization have for the performance of the technology or work system in this group?

:	: CK :	:	:	: SE / STM :	SS
It was not expected to do any more than the preceding system.	It was expected to do a *marginal* amount better (in quality, quantity, or cost)—perhaps 10% better.	It was expected to do *somewhat* more than its predecesor— about 50% better.	It was expected to do *a great deal* better in quantity, quality, or cost—perhaps twice as good.	It was expected to be a *major break-through* in this sort of work—possibly five times improvement.	This question is not applicable to this group.

If the technology or work system in the group is new or fairly new
(within the past four or five years) answer the following question.

12) To what extent is the new work system used in this group effective?   (To be effective the system must do what it was supposed to do—e.g., is it providing more and/or better output with the same number of people, or the same output with fewer human interventions?)

:	:	: CK / SE :	STM	:	: SS :
It is *very ineffective*— not operating nearly as well as expected.	It is *somewhat ineffective*— not operating quite as well as expected.	It is *effective*— operating as expected.	It is *very effective*— operating some- what better than expected.	It is *extremely effective*— operating far better than expected.	This question is not applicable to this group.

If the technology or work system in the group is new or fairly new
(within the past four or five years) answer the following question.

13) How long did it take to complete the start-up period for the installation of this system? That is, what was the period before the system operated as well as could be expected?

:	: STM / SE :	:	:	: CK	: SS :
It hasn't yet, but it's too soon to tell if things	It never did and probably never will operate well.	It took *8-30 months* before it operated well.	It took *6-8 months* before it operated well.	It took *3-5 months* before it operated well.	It took *1-2 months* before it operated well.	This question is not applicable to this group.

6. The responses to these scales have little basis in the descriptions you were provided but are included as examples of how these scales work.

APPENDIX B

Figure 14

CROSS-LAG CAUSAL ANALYSIS MATRICES

INSURANCE COMPANY TEST SITE DATA

High Technology (N groups = 29)

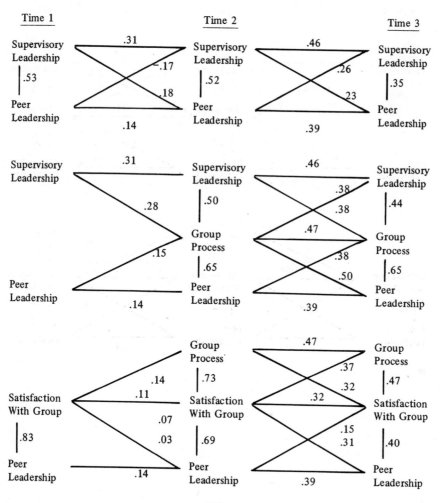

Figure 15

CROSS-LAG CAUSAL ANALYSIS MATRICES

INSURANCE COMPANY TEST SITE DATA

Low Technology (N groups = 22)

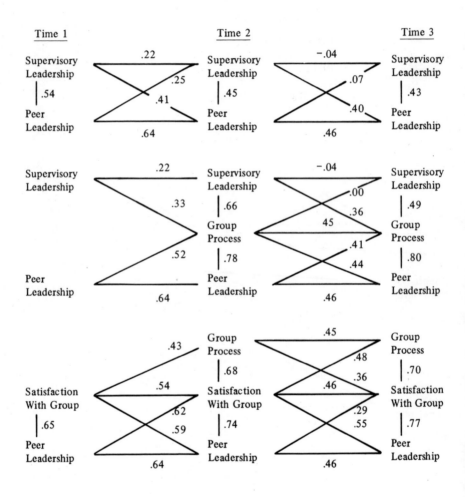

Figure 16

CROSS-LAG CAUSAL ANALYSIS MATRICES
REFINERY DATA

High Technology (N groups = 42)

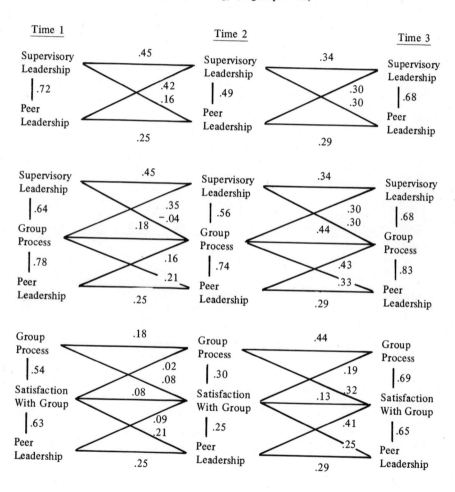

Figure 17

CROSS-LAG CAUSAL ANALYSIS MATRICES

REFINERY DATA

Low Technology (N groups = 26)

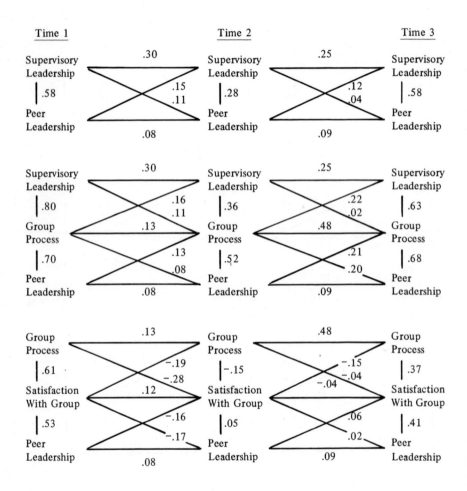

APPENDIX C

Table 17

MEAN SCORE DIFFERENCES ON DEPENDENT VARIABLES
BETWEEN HIGH-TECHNOLOGY, AND LOW-TECHNOLOGY GROUPS;
AND WITHIN GROUPS OVER TIME

Insurance Company Test Site Data – Maximization Procedure

Dependent Variables	High Technology			Low Technology			Differences	
	Means		diff.	Means		diff.		
	Time 3	Time 1	t_3-t_1	Time 3	Time 1	t_3-t_1	High$_1$-Low$_1$	High$_3$-Low$_3$
Supervisory Support	4.01	4.44	-.43**	4.22	4.63	-.41**	-.19#	-.21
Supervisory Goal Emphasis	3.88	4.02	-.14	4.07	4.19	-.12	-.17	-.19
Supervisory Work Facilitation	3.26	3.56	-.30*	3.54	3.58	-.04	-.02	-.28#
Supervisory Interaction Facilitation	3.66	3.68	-.02	3.92	3.83	.09	-.15	-.26#
Peer Support	3.81	4.12	-.31**	4.12	4.45	-.33**	-.33**	-.31**
Peer Goal Emphasis	3.47	3.55	.08	3.62	3.79	-.17#	-.24#	-.15
Peer Work Facilitation	3.32	3.24	-.06	3.41	3.20	.21	.04	-.09
Peer Interaction Facilitation	3.30	3.36	-.13	3.36	3.41	-.05	-.05	-.06
Group Process	3.47	3.60[a]		3.59	3.72[a]	-.13	-.12	-.12

#p < .10
*p < .05
**p < .01

[a]Group process was not measured at t_1, t_2 data used instead.

Table 18

MEAN SCORE DIFFERENCES ON DEPENDENT VARIABLES BETWEEN HIGH-TECHNOLOGY, AND LOW-TECHNOLOGY GROUPS; AND WITHIN GROUPS OVER TIME

Refinery Data – Maximization Procedure

Dependent Variables	High Technology			Low Technology			Differences	
	Means		diff.	Means		diff.		
	Time 3	Time 1	$t_3\text{-}t_1$	Time 3	Time 1	$t_3\text{-}t_1$	$High_1\text{-}Low_1$	$High_3\text{-}Low_3$
Supervisory Support	3.90	3.60	.30#	4.10	4.14	-.04	-.54*	-.20
Supervisory Goal Emphasis	3.93	3.81	.12	3.85	3.75	.10	.06	.08
Supervisory Work Facilitation	3.44	3.12	.32*	3.39	3.20	.19	-.08	.05
Supervisory Interaction Facilitation	3.59	3.35	.24	3.43	3.23	.20	.12	.16
Peer Support	3.81	3.65	.16	3.94	4.03	-.07	-.38*	-.13
Peer Goal Emphasis	3.46	3.50	-.04	3.48	3.44	.04	.06	-.02
Peer Work Facilitation	3.50	3.27	.39**	3.19	3.11	.08	.16	.31*
Peer Interaction Facilitation	3.40	3.19	.21*	2.87	2.68	.19	.51**	.53**
Group Process	3.67	3.59	.08	3.33	3.17	.16	.42**	.34*

#$p < .10$
*$p < .05$
**$p < .01$

REFERENCES

Allport, G. W. The historical background of modern social psychology. In G. Lindzey (Ed.), *Handbook of social psychology.* Vol. I. Cambridge, Mass.: Addison-Wesley, 1954.

American Foundation on Automation and Employment. *Automation and the middle manager.* New York: AFAE, 1966.

Andrews, F. M., Morgan, J. N., and Sonquist, J. A. *Multiple classification analysis.* Ann Arbor: Survey Research Center, Institute for Social Research, 1967.

Anshen, M. Managerial decisions. In J. T. Dunlop (Ed.), *Automation and technological change.* New York: The American Assembly, Columbia University, 1962.

Argyris, C. *Interpersonal competence and organizational effectiveness.* Homewood, Ill.: Irwin-Dorsey, 1962.

Bennis, W. G. *Changing organizations.* New York: McGraw-Hill, 1966.

Bereiter, C. Some persisting dilemmas in the measurement of change. In C. W. Harris (Ed.), *Problems in measuring change.* Madison: University of Wisconsin Press, 1967.

Blalock, H. M. *Causal inferences in nonexperimental research.* Chapel Hill: University of North Carolina Press, 1964.

Blau, P. M. *The dynamics of bureaucracy.* Chicago: The University of Chicago Press, 1955.

Blauner, R. *Alienation and freedom.* Chicago: The University of Chicago Press, 1964.

Blood, M. R., and Hulin, C. L. Alienation, environmental characteristics, and worker responses. *Journal of Applied Psychology,* 1967, *51*, 284-290.

Bowers, D. G. and Seashore, S. E. Predicting organizational effectiveness—with a four-factor theory of leadership. *Administrative Science Quarterly,* 1966, *11*, 238-263.

Bright, J. R. *Automation and management.* Boston: Division of Research, Harvard Business School, 1958.

Buckley, W. *Sociology and modern systems theory.* Englewood Cliffs, N. J.: Prentice-Hall, 1967.

Burns, T. and Stalker, G. M. *The management of innovation.* London: Tavistock Publications, 1961.

Campbell, D. T. From description to experimentation: Interpreting trends as quasi-experiments. In C. W. Harris (Ed.), *Problems in measuring change.* Madison: University of Wisconsin Press, 1967.

Chapple, E. D. and Sayles, L. R. *The measure of management: designing organizations for human effectiveness.* New York: Macmillan and Co., 1961.

Cohen, A. *Attitude change and social influence.* New York: Basic Books, 1964.

Converse, P. E. The nature of belief systems in mass publics. In D. Apter (Ed.), *Ideology and discontent.* Glencoe, Ill.: The Free Press, 1964.

Dubin, R. Supervision and productivity: empirical findings and theoretical considerations. In R. Dubin, G. C. Homans, F. C. Mann, and D. C. Miller (Eds.), *Leadership and productivity.* San Francisco: Chandler, 1965.

Duncan, O. D. Path Analysis: sociological examples. *American Journal of Sociology.* 1966, *72,* 1-16.

Emery, F. E. *Characteristics of socio-technical systems.* Unpublished manuscript. London: Tavistock Institute of Human Relations, January, 1959.

Faunce, W. A. Automation in the automobile industry. *American Sociological Review,* 1958, *23,* 401-407.

Faunce, W. A. *Problems of an industrial society.* New York: McGraw-Hill, 1968.

Festinger, L. *A theory of cognitive dissonance.* Stanford: Stanford University Press, 1957.

Friedmann, G. *The anatomy of work.* Glencoe: The Free Press, 1961.

Gouldner, A. W. Organizational analysis. In R. K. Merton, L. Broom and L. S. Cottrell (Eds.), *Sociology today,* New York: Basic Books, 1959.

Guest, R. H. *Organizational change: the effect of successful leadership.* Homewood, Ill.: The Dorsey Press, 1962.

Guttman, L. A general non-metric technique for finding the smallest coordinate spaces for a configuration of points, *Psychometrika,* 1968, *33,* 469-506.

Harbison, F. H., Kochling, E., Cassell, F. H. and Ruebmann, H. C. Steel management in two continents. *Management Science,* 1955, *2,* 31-39.

Harvey, E. Technology and the structure of organizations. *American Sociological Review,* 1968, *33,* 247-259.

Herbst, P. G. *Autonomous group functioning.* London: Tavistock Publications, 1962.

Hoos, I. R. *Automation in the office.* Washington, D.C.: Public Affairs Press, 1961.

Hulin, C. L. and Blood, M. R. Job enlargement, individual differences, and worker responses. *Psychological Bulletin,* 1968, *69,* 41-55.

Insko, C. A. *Theories of attitude change.* New York: Appleton-Century Crofts, 1967.

Jasinski, F. J. Horizontal and diagonal relationships in industrial plants. In C. R. Walker (Ed.), *Modern technology and civilization.* New York: McGraw-Hill, 1962.

Kahn, R. L. and Morse, N. C. The relationship of productivity to morale. *Journal of Social Issues,* 1951, *7,* 8-17.

Katz, D. and Kahn, R. L. *The social psychology of organizations.* New York: John Wiley & Sons, 1966.

Lewin, K. *Field theory in social science.* New York: Harper & Row, 1961.

Lieberman, S. The effects of changes in role in the attitudes of role occupants. *Human Relations,* 1956, *9,* 385-402.

Likert, R. *New patterns of management.* New York: McGraw-Hill, 1961.

Likert, R. *The human organization: its management and value.* New York: McGraw-Hill, 1967.

Lingoes, J. C. An IBM-7090 program for Guttman-Lingoes smallest space analysis, I. *Behavioral Science,* 1965, *10,* 183-184.

Lingoes, J. C. New computer developments in pattern analysis and nonmetric techniques. In *Uses of computers in psychological research.* Gauthier-Villors, Paris, 1966, 1-25.

Lingoes, J. C., Roskam, E.E.C.I., and Guttman, L. An empirical study of two multidimensional scaling algorithms, *Multivariate Behavioral Research,* 1969, *4.*

Mann, F. C. and Hoffman, L. R. *Automation and the worker.* New York: Henry Holt and Co., 1960.

Mann, F. C. and Williams, L. K. Organizational impact of white-collar automation. *Industrial Relations Research Association Proceedings.* 1959.

Mann, F. C. and Williams, L. K. Some effects of changing work environment in the office. *Journal of Social Issues,* 1962, *18,* 90-101.

Marenco, C. Gradualism, apathy, and suspicion in a French bank. In W. H. Scott (Ed.), *Office Automation,* Paris: OECD, 1965.

Marrow, A. J., Bowers, D. G., and Seashore, S. E. *Management by participation,* New York: Harper & Row, 1967.

Maslow, A. H. A theory of human motivation. *Psychological Bulletin,* 1943, *50,* 370-396.

McKinsey and Company. *Getting the most out of your computer.* New York: McKinsey and Company, 1958.

Mead, M. *New lives for old.* New York: Wm. Marrow & Company, 1956.

Morse, N. *Satisfactions in the white-collar job.* Ann Arbor, Michigan: Institute for Social Research, 1953.

Mueller, E. *Automation in an expanding economy.* Ann Arbor: Institute for Social Research, 1969.

Nangle, J. E. The effectiveness of communications in preparation for change in an insurance company. Unpublished doctoral dissertation, Michigan State University, 1961.

O'Connell, J. J. *Managing organizational innovation.* Homewood, Illinois: Richard D. Irwin, 1968.

Ogburn, W. F. In Allen, F. R., Hart, H., Miller, D. C., and Ogburn, W. F. (Eds.), *Technology and social change,* New York: Appleton-Century Crofts, 1957.

Ogburn, W. F. National policy and technology. In C. R. Walker (Ed.), *Modern technology and civilization.* New York: McGraw-Hill, 1962.

Pelz, D. C. and Andrews, F. M. Detecting causal priorities in panel study data. *American Sociological Review,* 1964, *29,* 836-848.

Perrow, C. A framework for the comparative analysis of organizations. *American Sociological Review,* 1967, *32,* 194-208.

Pugh, D. S. Modern organization theory: a psychological and sociological study. *Psychological Bulletin,* 1966, *66,* 235-251.

Rice, A. K. *Productivity and social organization: the Ahmedabad experiment.* London: Tavistock Publications, 1958.

Rice, A. K. *The enterprize and its environment: a system theory of management organization.* London: Tavistock Publications, 1963.

Rozelle, R. M. and Campbell, D. T. More plausible rival hypotheses in the cross-lagged panel correlation technique. *Psychological Bulletin,* 1969, *71,* 74-80.

Sayles, L. R. The change process in organizations: an applied anthropology analysis. *Human Organization,* 1962, *21,* 62-67.

Schachter, S., Willerman, B., Festinger, L., and Hyman, R. Emotional disruption and industrial productivity. *Journal of Applied Psychology,* 1961, *45,* 201-213.

Schein, E. H. Management development as a process of influence. *Industrial Management Review,* 1961, *2,* 59-77.

Schon, D. A. *Technology and change.* New York: Delta Books, 1967.

Scott, W. H. (Ed.)., *Office automation,* Paris: OECD, 1965.

Seashore, S. E. and Bowers, D. G. Durability of organizational change. *American Psychologist,* 1970, *25,* 227-233.

Simpson, R. L. Vertical and horizontal communication in formal organization. *Administrative Science Quarterly,* 1959, *4,* 188-196.

Tannenbaum, A. S. Personality change as a result of an experimental change of environmental conditions. *Journal of Abnormal and Social Psychology,* 1957, *55,* 404-406.

Taylor, J. C. Technology and its effects on work-related attitudes and behaviors: a review of the literature. Unpublished manuscript, Department of Psychology, The University of Michigan, 1968.

Taylor, J. C. Measuring sophistication of production technology: Background, development, and results. Duplicated research report to the United States Office of Naval Research, 1970, Contract N00014-67-A-0181-0013, NR170-719/7-29-69 (Code 452).

Taylor, J. C. An empirical examination of a four-factor theory of leadership using smallest space analysis. *Organizational Behavior and Human Performance.* 1971, *6,* 249-266.

Taylor, J. C. and Bowers, D. G. *The Survey of Organizations: Towards a machine-scored, standardized questionnaire instrument.* Ann Arbor: Institute for Social Research, 1971.

Thompson, J. D. *Organizations in action.* New York: McGraw-Hill, 1967.

Thompson, J. D. and Bates, F. L. Technology, organization, and administration. *Administrative Science Quarterly,* 1957, *2,* 325-343.

Thorsrud, E. Industrial democracy project in Norway, 1962-1968. Unpublished manuscript, Oslo: Work Research Institutes, 1968.

Touraine, A. An historical theory in the evolution of industrial skills. In C. R. Walker (Ed.), *Modern technology and civilization.* New York: McGraw-Hill, 1962.

Trist, E. L. and Bamforth, K. W. Some social and psychological consequences of the longwall method of coal getting. *Human Relations,* 1951, *4,* 3-38.

Trist, E. L., Higgin, G. W., Murray, H., and Pollock, A. B. *Organizational Choice.* London: Tavistock Publications, 1963.

Turner, A. N. and Lawrence, P. R. *Industrial jobs and the worker.* Cambridge: Harvard University Press, 1965.

Walker, C. R. *Toward the automatic factory.* New Haven: Yale University Press, 1957.

White, L. A. *The evolution of culture.* New York: McGraw-Hill, 1959.

Williams, L. K. and Williams, B. C. The impact of numerically controlled equipment on factory organization. *California Management Review,* 1964, *7,* 25-34.

Woodward, J. *Industrial organization: theory and practice.* Oxford: Oxford University Press, 1965.

Date Due

Nov 25 '73			

38-297